Data Science with
Julia

Data Science with Julia

By Paul D. McNicholas and Peter A. Tait

CRC Press
Taylor & Francis Group
Boca Raton London New York

CRC Press is an imprint of the
Taylor & Francis Group, an **informa** business

CRC Press
Taylor & Francis Group
6000 Broken Sound Parkway NW, Suite 300
Boca Raton, FL 33487-2742

Printed on acid-free paper
Version Date: 20191119

International Standard Book Number-13: 978-1-138-49998-0 (Paperback)

Library of Congress Cataloging-in-Publication Data

Names: McNicholas, Paul D., author. | Tait, Peter A., author.
Title: Data science with Julia / Paul D. McNicholas, Peter A. Tait.
Description: Boca Raton : Taylor & Francis, CRC Press, 2018. | Includes bibliographical references and index.
Identifiers: LCCN 2018025237 | ISBN 9781138499980 (pbk.)
Subjects: LCSH: Julia (Computer program language) | Data structures (Computer science)
Classification: LCC QA76.73.J85 M37 2018 | DDC 005.7/3--dc23
LC record available at https://lccn.loc.gov/2018025237

Visit the Taylor & Francis Web site at
http://www.taylorandfrancis.com

and the CRC Press Web site at
http://www.crcpress.com

For Oscar, who tries and tries.

PDM

To my son, Xavier,
Gettin' after it does pay off.

PAT

Contents

Foreword

The 21st century will probably be the century of the data revolution. Our numerical world is creating masses of data every day and the volume of generated data is increasing more and more (the number of produced numerical data is doubling every two years according to the most recent estimates). In such a context, data science is nowadays an unavoidable field for anyone interested in exploiting data. People may be interested in either understanding a phenomenon or in predicting the future behavior of this phenomenon.

To this end, it is important to have significant knowledge of both the rationale (the theory) behind data science techniques and their practical use on real-world data. Indeed, data science is a mix of data, statistical/machine learning methods and software. Software is actually the link between data and data science techniques. It allows the practitioner to load the data and apply techniques on it for analysis. It is therefore important to master at least one of the data science languages.

The choice of the software language(s) mainly depends on your background and the expected level of analysis. R and Python are probably the two most popular languages for data science. On the one hand, R has been made by statisticians... mostly for statisticians! It is, however, an excellent tool for data science since the most recent statistical learning techniques are provided on the R platform (named CRAN). Using R is probably the best way to be directly connected to current research in statistics and data science through the packages provided by researchers. Python is, on the other hand, an actual computer science language (with all appropriate formal aspects) for which some advanced libraries for data science exist. In this context, the Julia language has the great advantage to permit users to interact with both R and Python (but also C, Fortran, etc.), within a software language designed for efficient and parallel numerical computing while keeping a high level of human readability.

Professor Paul McNicholas and Peter Tait propose in this book to learn both fundamental aspects of data science: theory and application. First, the book will provide you with the significant elements to understand the mathematical aspects behind the most used data science techniques. The book will also allow you to discover advanced recent techniques, such as probabilistic principal components analysis (PPCA), mixtures of PPCAs, and gradient boosting. In addition, the book will ask you to dive into the Julia language such that you directly apply the learned techniques on concrete examples. This is, in my opinion, the most efficient way to learn such an applied science. In addition, the focus made by this book on the Julia language is a great choice because of the numerous qualities of this language regarding data science practice. These include ease of learning for people familiar with R or Python, nice syntax, easy code debugging, the speed of the compiled language, and code reuse.

Both authors have extensive experience in data science. Professor Paul McNicholas is Canada Research Chair in Computational Statistics at McMaster University and Director of the MacDATA Institute of the same university. In his already prolific career, McNicholas has made important contributions to statistical learning. More precisely, his research is mainly focused on model-based learning with high-dimensional and skew-distributed data. He is also a researcher deeply involved in the spreading of research products through his numerous contributions to the R software with packages. Peter Tait is currently a Ph.D. student but, before returning to academia, he had a professional life dedicated to data science in industry. His strong knowledge of the needs of industry regarding data science problems was really an asset for the book.

This book is a great way to both start learning data science through the promising Julia language and to become an efficient data scientist.

Professor Charles Bouveyron
Professor of Statistics
INRIA Chair in Data Science
Université Côte d'Azur
Nice, France

Preface

This is a book for people who want to learn about the Julia language with a view to using it for data science. Some effort has gone into making this book suitable for someone who has familiarity with the R software and wants to learn about Julia. However, prior knowledge of R is not a requirement. While this book is not intended as a textbook for a course, some may find it a useful book to follow for a course that introduces statistics or data science students to Julia. It is our sincere hope that students, researchers and data scientists in general, who wish to learn Julia, will find this book beneficial.

More than twenty years have passed since the term data science was described by Dr. Chikio Hayashi in response to a question at a meeting of the International Federation of Classification Societies (Hayashi, 1998). Indeed, while the term data science has only gained notoriety over the past few years, much of the work it describes has been practiced for far longer. Furthermore, whatever the precise meaning of the term, there is no doubt that data science is important across virtually all areas of endeavour. This book is born out of a mixture of experiences all of which led to the conclusion that the use of Julia, as a language for data science, should be encouraged.

First, part of the motivation to write this book came from experience gained trying to teach material in data science without the benefit of a relatively easily understood base language that is effective for actually writing code. Secondly, there is the practical, and related, matter of writing efficient code while also having access to excellent code written by other researchers. This, of course, is the major advantage of R, where many researchers have contributed packages — sometimes based on code written in another language such as C or Fortran — for a wide variety of statistics and data science tasks. As we illustrate in this book, it is straightforward to call R from Julia and to thereby access whatever R packages are needed. Access to R packages and a growing selection of Julia packages, together with an accessible, intuitive, and highly efficient base language, makes Julia a formidable platform for data science. This book is not intended as an exhaustive introduction to data science. In fact, this book is far from an exhaustive introduction to data science. There are many very good books that one can consult to learn about different aspects of data science (e.g., Bishop, 2006; Hastie et al., 2009; Schutt, 2013; White, 2015; Efron and Hastie, 2016), but this book is primarily about Julia. Nevertheless, several important topics in data science are covered. These include data visualization, supervised learning, and unsupervised learning. When discussing supervised learning, we place some focus on gradient boosting — a machine learning technique — because we have found this approach very effective in applications. However, for unsupervised learning, we take a more statistical approach and place some focus on the use of probabilistic principal components analyzers and a mixture thereof.

This monograph is laid out to progress naturally. In Chapter 1, we discuss data science and provide some historical context. Julia is also introduced as well as details of the packages and datasets used herein. Chapters 2 and 3 cover the basics of the Julia language was well as how to work with data in Julia. After that (Chapter 4), a crucially important topic in data science is discussed: visualization. The book continues with selected techniques in supervised (Chapter 5) and unsupervised learning (Chapter 6), before concluding with details of how to call R functions from within Julia (Chapter 7). This last chapter also provides further examples of mixture model-based clustering as well as an example that uses random forests. Some appendices are included to provide readers with some relevant mathematics, Julia performance tips and a list of useful linear algebra functions in Julia.

There is a large volume of Julia code throughout this book, which is intended to help the reader gain familiarity with the language. We strongly encourage readers to run the code for themselves and play around with it. To make the code as easy as possible to work with, we have interlaced it with comments. As readers begin to get to grips with Julia, we encourage them to supplement or replace our comments with their own. For the reader's convenience, all of the code from this book is available on GitHub: github.com/paTait/dswj.

We are most grateful to David Grubbs of the Taylor & Francis Group for his support in this endeavour. His geniality and professionalism are always very much appreciated. Special thanks to

Professor Charles Bouveyron for kindly agreeing to lend his expertise in the form of a wonderful Foreword to this book. Thanks also to Dr. Joseph Kang and an anonymous reviewer for their very helpful comments and suggestions. McNicholas is thankful to Eamonn Mullins and Dr. Myra O'Regan for providing him with a solid foundation for data science during his time as an undergraduate student. Dr. Sharon McNicholas read a draft of this book and provided some very helpful feedback for which we are most grateful.

A final word of thanks goes to our respective families; without their patience and support, this book would not have come to fruition.

Paul D. McNicholas and Peter A. Tait
Hamilton, Ontatio

About the Authors

Paul D. McNicholas is the Canada Research Chair in Computational Statistics at McMaster University, where he is a Professor and University Scholar in the Department of Mathematics and Statistics as well as Director of the MacDATA Institute. He has published extensively in computational statistics, with the vast majority of his work focusing on mixture model-based clustering. He is one of the leaders in this field and recently published a monograph devoted to the topic (*Mixture Model-Based Classification*; Chapman & Hall/CRC Press, 2016). He is a Senior Member of the IEEE and a Member of the College of the Royal Society of Canada.

Peter A. Tait is a Ph.D. student at the School of Computational Science and Engineering at McMaster University. His research interests span multivariate and computational statistics. Prior to returning to academia, he worked as a data scientist in the software industry, where he gained extensive practical experience.

## CHAPTER 1

Introduction

D ATA SCIENCE is discussed and some important connections, and contrasts, are drawn between statistics and data science. A brief discussion of big data is provided, the Julia language is briefly introduced, and all Julia packages used in this monograph are listed together with their respective version numbers. The same is done for the, albeit smaller number of, R packages used herein. Providing such details about the packages used helps ensure that the analyses illustrated herein can be reproduced. The datasets used in this monograph are also listed, along with some descriptive characteristics and their respective sources. Finally, the contents of this monograph are outlined.

1.1 DATA SCIENCE

What is data science? It is an interesting question and one without a widely accepted answer. Herein, we take a broad view that data science encompasses all work related to data. While this includes data analysis, it also takes in a host of other topics such as data cleaning, data curation, data ethics, research data management, etc. This monograph discusses some of those aspects of data science that are commonly handled in Julia, and similar software; hence, its title.

The place of statistics within the pantheon of data science is a topic on which much has been written. While statistics is certainly a very important part of data science, statistics should not be taken as synonymous with data science. Much has been written about the relationship between data science and statistics. On the one extreme, some might view data science — and data analysis, in particular — as a retrogression of statistics; yet, on the other extreme, some may argue that data science is a manifestation of what statistics was always meant to be. In reality, it is probably an error to try to compare statistics and data science as if they were alternatives. Herein, we consider that statistics plays a crucial role in data analysis, or data analytics, which in turn is a crucial part of the data science mosaic.

Contrasting data analysis and mathematical statistics, Hayashi (1998) writes:

> ... mathematical statistics have been prone to be removed from reality. On the other hand, the method of data analysis has developed in the fields disregarded by mathematical statistics and has given useful results to solve complicated problems based on mathematico-statistical methods (which are not always based on statistical inference but rather are descriptive).

The views expressed by Hayashi (1998) are not altogether different from more recent observations that, insofar as analysis is concerned, data science tends to focus on prediction, while statistics has focused on modelling and inference. That is not to say that prediction is not a part of inference but rather that prediction is a part, and not the goal, of inference. We shall return to this theme, i.e., inference versus prediction, several times within this monograph.

Breiman (2001b) writes incisively about two cultures in statistical modelling, and this work is wonderfully summarized in the first few lines of its abstract:

> There are two cultures in the use of statistical modeling to reach conclusions from data. One assumes that the data are generated by a given stochastic data model. The other uses algorithmic models and treats the data mechanism as unknown. The statistical community has been committed to the almost exclusive use of data models. This commitment has led to irrelevant theory, questionable conclusions, and has kept statisticians from working on a large range of interesting current problems.

The viewpoint articulated here leans towards a view of data analysis as, at least partly, arising out of one culture in statistical modelling.

In a very interesting contribution, Cleveland (2001) outlines a blueprint for a university department, with knock-on implications for *curricula*. Interestingly, he casts data science as an "altered field" — based on statistics being the base, i.e., unaltered, field. One fundamental alteration concerns the role of computing:

> One outcome of the plan is that computer science joins mathematics as an area of competency for the field of data science. This enlarges the intellectual foundations. It implies partnerships with computer scientists just as there are now partnerships with mathematicians.

Writing now, as we are 17 years later, it is certainly true that computing has become far more important to the field of statistics and is central to data science. Cleveland (2001) also presents two contrasting views of data science:

> A very limited view of data science is that it is practiced by statisticians. The wide view is that data science is practiced by statisticians and subject matter analysts alike, blurring exactly who is and who is not a statistician.

Certainly, the wider view is much closer to what has been observed in the intervening years. However, there are those who can claim to be data scientists but may consider themselves neither statisticians nor subject matter experts, e.g., computer scientists or librarians and other data curators. It is noteworthy that there is a growing body of work on how to introduce data science into *curricula* in statistics and other disciplines (see, e.g., Hardin et al., 2015).

One fascinating feature of data science is the extent to which work in the area has penetrated into the popular conscience, and media, in a way that statistics has not. For example, Press (2013) gives a brief history of data science, running from Tukey (1962) to Davenport and Patil (2012) — the title of the latter declares data scientist the "sexiest job of the 21st century"! At the start of this timeline is the prescient paper by Tukey (1962) who, amongst many other points, outlines how his view of his own work moved away from that of a statistician:

> For a long time I have thought I was a statistician, interested in inferences from the particular to the general. ...All in all, I have come to feel that my central interest is in *data analysis*, which I take to include, among other things: procedures for analyzing data, techniques for interpreting the results of such procedures, ways of planning the gathering of data to make its analysis easier, more precise or more accurate, and all the machinery and results of (mathematical) statistics which apply to analyzing data.

The wide range of views on data science, data analytics and statistics thus far reviewed should serve to convince the reader that there are differences of opinion about the relationship between these disciplines. While some might argue that data science, in some sense, *is* statistics, there seems to be a general consensus that the two are not synonymous. Despite the modern views expounded by Tukey (1962) and others, we think it is fair to say that much research work within the field of statistics remains mathematically focused. While it may seem bizarre to some readers, there are still statistics researchers who place more value in an ability to derive the mth moment of some obscure distribution than in an ability to actually analyze real data. This is not to denigrate mathematical statistics or to downplay the crucial role it plays within the field of statistics; rather, to emphasize that there are some who value ability in mathematical statistics far more than competence in data analysis. Of course, there are others who regard an ability to analyze data as a *sine qua non* for anyone who would refer to themselves as a statistician. While the proportion of people holding the latter view may be growing, the rate of growth seems insufficient to suggest that we will shortly arrive at a point where a statistician can automatically be assumed capable of analyzing data.

This latter point may help to explain why the terms data science, data scientist and data analyst are important. The former describes a field of study devoted to data, while the latter two describe people who are capable of working with data. While it is true that there are many statisticians who may consider themselves data analysts, it is also true that there are many data analysts who are not statisticians.

1.2 BIG DATA

Along with rapidly increasing interest in data science has come the popularization of the term big data. Similar to the term data science, big data has no universally understood meaning. Puts et al. (2015) and others have described big data in terms of words that begin with the letter V: volume, variety, and velocity. Collectively, these can be thought of as the three Vs that define big data; however, other V words have been proposed as part of such a definition, e.g., veracity, and alternative definitions have also been proposed. Furthermore, the precise meaning of these V words is unclear. For instance, volume can be taken as referring to the overall quantity of data or the number of dimensions (i.e., variables) in the dataset. Variety can be taken to mean that data come from different sources or that the variables are of different types (such as interval, nominal, ordinal, binned, text, etc.). The precise meaning of velocity is perhaps less ambiguous in that it is usually taken to mean that data come in a stream. The word veracity, when included, is taken as indicative of the extent to which the data are reliable, trustworthy, or accurate. Interestingly, within such three (or more) Vs definitions, it is unclear how many Vs must be present for data to be considered big data.

The buzz attached to the term big data has perhaps led to some attempts to re-brand well-established data as somehow big. For instance, very large databases have existed for many years but, in some instances, there has been a push to refer to what might otherwise be called administrative data as big data. Interestingly, Puts et al. (2015) draw a clear distinction between big data and administrative data:

> Having gained such experience editing large administrative data sets, we felt ready to process Big Data. However, we soon found out we were unprepared for the task.

Of course, the precise meaning of the term big data is less important than knowing how to tackle big data and other data types. Further to this point, we think it is a mistake to put big data on a pedestal and hail it as *the* challenging data. In reality there are many challenging datasets that do not fit within a definition of big data, e.g., situations where there is very little data are notoriously difficult. The view that data science is essentially the study of big data has also been expounded and, in the interest of completeness, deserves mention here. It is also important to clarify that we reject this view out of hand and consider big data, whatever it may be, as just one of the challenges faced in data analysis or, more broadly, in data science. Hopefully, this section has provided some useful context for what big data is. The term big data, however, will not be revisited within this monograph, save for the References.

1.3 JULIA

The Julia software (Bezansony et al., 2017) has tremendous potential for data science. Its syntax is familiar to anyone who has programmed in R (R Core Team, 2018) or Python (van Rossum, 1995), and it is quite easy to learn. Being a dynamic programming language specifically designed for numerical computing, software written in Julia can attain a level of performance nearing that of statically-compiled languages like C and Fortran. Julia integrates easily into existing data science pipelines because of its superb language interoperability, providing programmers with the ability to interact with R, Python, C and many other languages just by loading a Julia package. It uses computational resources very effectively so that sophisticated algorithms perform extremely well on a wide variety of hardware. Julia has been designed from the ground up to take advantage of the parallelism built into modern computer hardware and the distributed environments available for software deployment. This is not the case for most competing data science languages. One additional benefit of writing software in Julia is how human-readable the code is. Because high-performance code does not require vectorization or other obfuscating mechanisms, Julia code is clear and straightforward to read, even months after being written. This can be a true benefit to data scientists working on large, long-term projects.

1.4 JULIA AND R PACKAGES

Many packages for the Julia software are used in this monograph as well as a few packages for the R software. Details of the respective packages are given in Appendix A. Note that Julia version 1.0.1 is used herein.

1.5 DATASETS

1.5.1 Overview

The datasets used for illustration in this monograph are summarized in Table 1.1 and discussed in the following sections. Note that datasets in Table 1.1 refers to the datasets package which is part of R, MASS refers to the MASS package (Venables and Ripley, 2002) for R, and mixture refers to the mixture package (Browne and McNicholas, 2014) for R. For each real dataset, we clarify whether or not the data have been pre-cleaned. Note that, for our purposes, it is sufficient to take pre-cleaned to mean that the data are provided, at the source, in a form that is ready to analyze.

1.5.2 Beer Data

The beer dataset is available from www.kaggle.com and contains data on 75,000 home-brewed beers. The 15 variables in the beer data are described in Table 1.2. Note that these data are pre-cleaned.

Table 1.1 The datasets used herein, with the number of samples, dimensionality (i.e., number of variables), number of classes, and source.

Name	Samples	Dimensionality	Classes	Source
beer	75,000	15	–	www.kaggle.com
coffee	43	12	2	pgmm
crabs	200	5	2 or 4	MASS
food	126	60	–	www.kaggle.com
iris	150	4	3	datasets
x2	300	2	3	mixture

Table 1.2 Variables for the beer dataset.

Variable	Description
ABV	Alcohol by volume.
BoilGravity	Specific gravity of wort before boil.
BoilSize	Fluid at beginning of boil.
BoilTime	Time the wort is boiled.
Colour	Colour from light (1) to dark (40).
Efficiency	Beer mask extraction efficiency.
FG	Specific gravity of wort after fermentation.
IBU	International bittering units.
Mash thickness	Amount of water per pound of grain.
OG	Specific gravity of wort before fermentation.
PitchRate	Yeast added to the fermentor per gravity unit.
PrimaryTemp	Temperature at fermentation stage.
PrimingMethod	Type of sugar used for priming.
PrimingAmount	Amount of sugar used for priming.
SugarScale	Concentration of dissolved solids in wort.

1.5.3 Coffee Data

Streuli (1973) reports on the chemical composition of 43 coffee samples collected from 29 different countries. Each sample is either of the Arabica or Robusta species, and 12 of the associated chemical constituents are available as the coffee data in pgmm (Table 1.3). One interesting feature of the coffee data — and one which has been previously noted (e.g., Andrews and McNicholas, 2014; McNicholas, 2016a) — is that Fat and Caffeine perfectly separate the Arabica and Robusta samples (Figure 1.1). Note that these data are also pre-cleaned.

Table 1.3 The 12 chemical constituents given in the `coffee` data.

Water	Bean Weight	Extract Yield
pH Value	Free Acid	Mineral Content
Fat	Caffeine	Trigonelline
Chlorogenic Acid	Neochlorogenic Acid	Isochlorogenic Acid

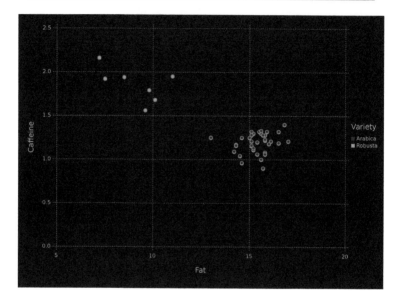

Figure 1.1 Scatterplot of fat versus caffeine, coloured by variety, for the `coffee` data.

1.5.4 Leptograpsus Crabs Data

The `crabs` data are available in the `MASS` library for R. Therein are five morphological measurements (Table 1.4) on two species of crabs (blue and orange), further separated into two sexes. The variables are highly correlated (e.g., Figures 1.2–1.4). As noted in Table 1.1, these data can be regarded as having two or four classes. The two classes can be taken as corresponding to either species (Figure 1.2) or sex (Figure 1.3), and the four-class solution considers both species and sex (Figure 1.4). Regardless of how the classes are broken down, these data represent a difficult clustering problem — see Figures 1.2–1.4 and consult McNicholas (2016a) for discussion. These data are pre-cleaned.

Table 1.4 The five morphological measurements given in the crabs data, all measured in mm.

Frontal lobe size	Rear width
Carapace length	Carapace width
Body depth	

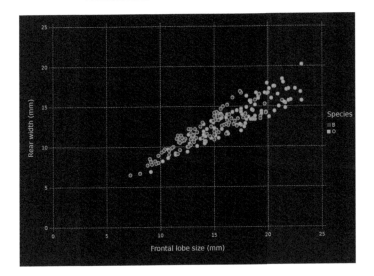

Figure 1.2 Scatterplot for rear width versus frontal lobe, for the crabs data, coloured by species.

1.5.5 Food Preferences Data

The food dataset is available from www.kaggle.com and contains data on food choices and preferences for 126 students from Mercyhurst University in Erie, Pennsylvania. The variables in the food data are described in Tables B.1–B.1 (Appendix B). Note that these data are raw, i.e., not pre-cleaned.

1.5.6 x2 Data

The x2 data are available in the mixture library for R. The data consist of 300 bivariate points coming, in equal proportions, from one of three Gaussian components (Figure 1.5). These data have been used to demonstrate clustering techniques (see, e.g., McNicholas, 2016a).

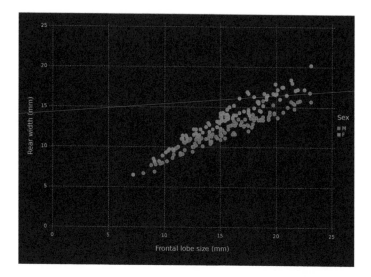

Figure 1.3 Scatterplot for rear width versus frontal lobe, for the crabs data, coloured by sex.

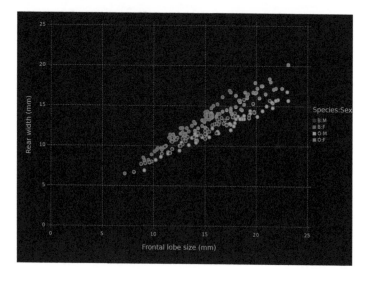

Figure 1.4 Scatterplot for rear width versus frontal lobe, for the crabs data, coloured by species and sex.

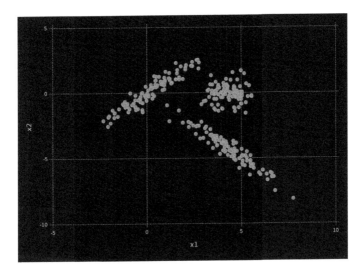

Figure 1.5 Scatterplot depicting the x2 data.

1.5.7 Iris Data

The `iris` data (Anderson, 1935) are available in the `datasets` library for R. The data consists of four measurements (Table 1.5) for 150 irises, made up of 50 each from the species *setosa*, *versicolor*, and *virginica*. These data are pre-cleaned.

Table 1.5 The four measurements taken for the iris data, all measured in cm.

Sepal length	Sepal width
Petal length	Petal width

1.6 OUTLINE OF THE CONTENTS OF THIS MONO-GRAPH

The contents of this monograph proceed as follows. Several core Julia topics are described in Chapter 2, including variable names, types, data structures, control flow, and functions. Various tools needed for working with, or handling, data are discussed in Chapters 3, including dataframes and input-output (IO). Data visualization, a crucially important topic in data science, is discussed in Chapter 4. Selected supervised learning techniques are discussed in Chapter 5, including K-nearest neighbours classification, classification and regression trees, and gradient boosting. Unsupervised learning follows (Chapter 6), where k-means clustering is discussed along with probabilistic principal components analyzers and mixtures thereof. The final chapter (Chapter 7) draws together Julia with R, principally by explaining how R code can be run from within Julia, thereby allowing R functions to be called from within Julia. Chapter 7 also illustrates further supervised and unsupervised learning techniques, building on the contents of Chapters 5 and 6. Five appendices are included. As already mentioned, the Julia and R packages used herein are detailed in Appendix A, and variables in the food data are described in Appendix B. Appendix C provides details of some mathematics that are useful for understanding some of the topics covered in Chapter 6. In Appendix D, we provide some helpful performance tips for coding in Julia and, in Appendix E, a list of linear algebra functions in Julia is provided.

## CHAPTER 2

Core Julia

T HE PURPOSE of this chapter is to expose the reader to the core parts of the Julia language. While it is intended to serve as an introduction to the language, it should also prove useful as a reference as the reader moves through the remainder of the book. The topics covered can be considered under six headings: variable names, operators, types, data structures, control flow, and functions. Furthermore, the chapter is laid out so that the contents build in a cumulative fashion, e.g., by the time one reaches the material on functions, all of the other material is being, or could usefully be, used. Accordingly, the material in this chapter is best read in order upon first reading.

2.1 VARIABLE NAMES

In Julia, variable names are associated with a particular value. This value is stored so it can be used in future computations or other operations. Variable names are case sensitive and Unicode names (in UTF-8 encoding) may be used. Specifically, Unicode math symbols can be entered by using the L^AT_EX2_ε symbol followed by a tab. In general, variable names must begin with a letter (uppercase or lowercase), an underscore _ or a Unicode code point > 00A0, and subsequent characters can include the ! symbol, digits and other Unicode points. It is also possible to redefine constants (e.g., π); however, built-in statements cannot be used as variable names (example below).

```
## Some playing around with variable names

## These are all correct
z = 100
```

```
y = 10.0
s = "Data Science"
ϑ = 5.0
μ = 1.2
datascience = true
data_science = true

## These are incorrect and will return an error
if = true
else = 1.5
```

The recommended convention is that variable names are typed in lowercase letters and words are separated by an underscore. In general, this convention will be followed herein.

2.2   OPERATORS

In programming languages, operators are symbols used to do specific mathematical, logical or relational operations. Julia operators will be familiar to R and Python users. There are four main categories of operators in Julia: arithmetic, updating, numeric comparison, and bitwise. Examples of the first three are given in the following code block.

```
x = 2
y = 3
z = 4

## Arithmetic operators
x + y
# 5

x^y
# 8

## Updating operators
x += 2
# 4

y -= 2
# 1

z *= 2
# 8

## Numeric Comparison
x == y
# false

x != y
# true

x <= z
# true
```

When constructing expressions with multiple operators, the order in which these operators are applied to the expression is known as operator precedence. Operator precedence is an important practical consideration and can lead to unexpected results. The following code block uses the variables instantiated above to evaluate three different expressions. The first one is evaluated solely based on the operator precedence. The following two are forced to evaluate in specific ways based on the parentheses () included in the expression. Parentheses are evaluated first in the precedence hierarchy. We recommend programming expressions with parentheses to minimize bugs and improve code clarity.

```
## Operator precedence
x*y+z^2
# 68

x*(y+(z^2))
# 260

(x*y)+(z^2)
# 68
```

2.3 TYPES

2.3.1 Numeric

Numeric literals are representations of numbers in code. Numeric primitives are representations of numbers as objects in memory. Julia has many primitive numeric types, e.g., Int32, Int64, Float32, and Float64. Julia offers full support for real and complex numbers. The internal variable Sys.WORD_SIZE displays the architecture type of the computer (e.g., 32 bit or 64 bit). The minimum and maximum values of numeric primitives can be displayed with the functions typemin() and typemax(), respectively. They take the name of numeric primitives as an argument and are detailed in the following code block. The default size of the primitive depends on the type of computer architecture.

```
## Computer's architecture type
Sys.WORD_SIZE
# 64

## Size of the default primitive
typemax(Int)
# 9223372036854775807
```

```
## Size of a specific primitive
## same as the default
typemax(Int64)
# 9223372036854775807

# Note that the above results are machine-specific.
```

A signed type can hold positive or negative integers, and uses the leftmost bit to identify the sign of the integer (e.g., Int64). An unsigned type can hold positive values only and stores larger values by using the leading bit as part of the value (e.g., UInt128). Boolean values are 8-bit integers, with false being 0 and true being 1. When doing arithmetic with integers, occasionally one will encounter overflow errors. This occurs when the result of an arithmetic expression is a value outside the representable range of the numeric primitive being used. This can happen if the result is larger or smaller than its allowable size. Examples are given in the code block below. If this is a possibility in a particular application, consider using unsigned integers or arbitrary precision integers, available in Julia as the BigInt type.

```
## Some examples of the Int type

# Integers
literal_int = 1
println("typeof(literal_int): ", typeof(literal_int))
# typeof(literal_int): Int64

# Boolean values
x = Bool(0)
y = Bool(1)

## Integer overflow error
x = typemax(Int64)
# 9223372036854775808
x += 1
# -9223372036854775807
x == typemax(Int64)
#false

## Integer underflow error
x = typemin(Int64)
# -9223372036854775808
x -= 1
# 9223372036854775807
x == typemin(Int64)
#false
```

2.3.2 Floats

Floats are similar to scientific notation. They are made up of three components: a signed integer whose length determines the precision (the significand); the base used to represent the number (usually base 10); and a signed integer that changes the magnitude of the floating point number (the exponent). The value of a float is determined by multiplying the significand by the base raised to the power of the exponent. Float64 literals are distinguished by having an e before the power, and can be defined in hexadecimal. Float32 literals are distinguished by having an f in place of the e. There are three Float64 values that do not occur on the real line:

1. Inf, positive infinity: a value larger than all finite floating point numbers, equal to itself, and greater than every other floating point value but NaN.

2. -Inf, negative infinity: a value less than all finite floating point numbers, equal to itself, and less than every other floating point value but NaN.

3. NaN, not a number: a value not equal to any floating point value, and not ==, < or > than any floating point value, including itself.

It is good practice to check for NaN, Inf and -Inf values in floating point operations. The following code block gives some examples of how to do this.

```
## Some examples of floats

x1 = 1.0
x64 = 15e-5
x32 = 2.5f-4

typeof(x32)
# Float32

## digit separation using an _
9.2_4 == 9.24
# true

isnan(0/0)
# true
isinf(1/0)
# true
isinf(-11/0)
# true

y1 = 2*3
```

```
# 6
isnan(y1)
# false
isinf(y1)
# false

y2 = 2/0
# Inf
isnan(y2)
# false
isinf(y2)
# true
```

The IEEE 754 standard (Zuras et al., 2008) sets out the technical standard for floating point arithmetic. The Julia `Float` type and all operations performed on it adhere to this standard. Representations of `Float64` numbers are not evenly spaced, with more occurring closer to zero on the real number line. This is due to machine epsilon, an upper-bound on the rounding error in floating point arithmetic. It is defined to be the smallest value of z such that $1+z \neq 1$. In Julia, the value of epsilon for a particular machine can be found via the `eps()` function. The spacing between floating point numbers and the value of machine epsilon is important to understand because it can help avoid certain types of errors. Integer overflow errors have been mentioned, but there are also float underflow errors, which occur when the result of a calculation is smaller than machine epsilon or when numbers of similar precision are subtracted. We give more details in Appendix D, and readers are directed to Higham (2002) if further reading on the topic is desired.

```
## Some examples of machine epsilon

eps()
# 2.220446049250313e-16

## spacing between a floating point number x and adjacent number is
## at most eps * abs(x)

n1 =[1e-25, 1e-5, 1., 1e5, 1e25]

for i in n1
  println( *(i, eps() ))
end
# 2.2204460492503132e-41
# 2.2204460492503133e-21
# 2.220446049250313e-16
# 2.220446049250313e-11
# 2.2204460492503133e9
```

Note that, as is common in scientific notation,

$$2.2204460492503132e-41$$

represents
$$2.2204460492503132 \times 10^{-41}.$$

2.3.3 Strings

In Julia, a string is a sequence of Unicode code points, using UTF-8 encoding. The first 128 Unicode characters are the ASCII characters. Characters in strings have an index value within the string. It is worth noting that Julia indices start at position 1, similar to R but different to Python, which starts its indices at position 0. The keyword end can be used to represent the last index. Herein, we will deal with ASCII characters only. Note that String is the built-in type for strings and string literals, and Char is the built-in type used to represent single characters. In fact, Char is a numeric value representing a Unicode code point. The value of a string cannot be changed, i.e., strings are immutable, and a new string must be built from another string. Strings are defined by double or triple quotes.

```
## String examples

s1 = "Hi"
# "Hi"

s2 = """I have a "quote" character"""
#  "I have a \"quote\" character"
```

Strings can be sliced using range indexing, e.g., my_string[4:6] would return a substring of my_string containing the 4th, 5th and 6th characters of my_string. Concatenation can be done in two ways: using the string() function or with the * operator. Note this is a somewhat unusual feature of Julia — many other languages use + to perform concatenation. String interpolation takes place when a string literal is defined with a variable inside its instantiation. The variable is prepended with $. By using variables inside the string's definition, complex strings can be built in a readable form, without multiple string multiplications.

```
## Some examples of strings
str = "Data science is fun!"
```

```
str[1]
# 'D'

str[end]
#'!'

## Slicing
str[4:7]
# "a sc"

str[end-3:end]
# "fun!"

## Concatenation
string(str, " Sure is :)")
#"Data science is fun! Sure is :)"

str * " Sure is :)"
# "Data science is fun! Sure is :)"

## Interpolation
"1 + 2 = $(1 + 2)"
#"1 + 2 = 3"

word1 = "Julia"
word2 = "data"
word3 = "science"
"$word1 is great for $word2 $word3"
#"Julia is great for data science"
```

Strings can be compared lexicographically using comparison operators, e.g., ==, >, etc. Lexicographical comparison involves sequentially comparing string elements with the same position, until one pair of elements falsifies the comparison, or the end of the string is reached. Some useful string functions are:

- findfirst(pat, str) returns the indices of the characters in the string str matching the pattern pat.

- occursin(substr, str) returns true/false depending on the presence/absence of substr in str.

- repeat(str, n) generates a new string that is the original string str repeated n times.

- length(str) returns the number of characters in the string str.

- replace(str, ptn => rep) searches string str for the pattern ptn and, if it is present, replaces it with rep.

Julia fully supports regular expressions (regexes). Regexes in Julia are fully Perl compatible and are used to hunt for patterns in string data. They are defined as strings with a leading r outside the quotes. Regular expressions are commonly used with the following functions:

- occursin(regex, str) returns true/false if the regex has a match in the string str.

- match(regex, str) returns the first match of regex in the string. If there is no match, it returns the special Julia value nothing.

- eachmatch(regex, str) returns all the matches of regex in the string str as an array.

Regexes are a very powerful programming tool for working with text data. However, an in-depth discussion of them is beyond the scope of this book, and interested readers are encouraged to consult Friedl (2006) for further details.

```
## Lexicographical comparison
s1 = "abcd"
s2 = "abce"

s1 == s2
# false

s1 < s2
# true

s1 > s2
#false

## String functions
str = "Data science is fun!"

findfirst("Data", str)
# 1:4

occursin("ata", str )
# true

replace(str, "fun" => "great")
# "Data science is great!"

## Regular expressions
## match alpha-numeric characters at the start of the str
occursin(r"^[a-zA-Z0-9]", str)
# true

## match alpha-numeric characters at the end of the str
occursin(r"[a-zA-Z0-9]$", str)
# false

## matches the first non-alpha-numeric character in the string
match(r"[^a-zA-Z0-9]", str)
```

```
#RegexMatch(" ")

## matches all the non-alpha-numeric characters in the string
collect(eachmatch(r"[^a-zA-Z0-9]", str))
#4-element Array{RegexMatch,1}:
# RegexMatch(" ")
# RegexMatch(" ")
# RegexMatch(" ")
# RegexMatch("!")
```

2.3.4 Tuples

Tuples are a Julia type. They are an abstraction of function arguments without the function. The arguments are defined in a specific order and have well-defined types. Tuples can have any number of parameters, and they do not have field names. Fields are accessed by their index, and tuples are defined using brackets () and commas. A very useful feature of tuples in Julia is that each element of a tuple can have its own type. Variable values can be assigned directly from a tuple where the value of each variable corresponds to a value in the tuple.

```
## A tuple comprising only floats
tup1 = (3.0, 9.1, 0.8, 1.9)
tup1
# (3.0, 9.1, 0.8, 1.9)
typeof(tup1)
# NTuple{4,Float64}

## A tuple comprising strings and floats
tup2 = ("Data", 2.5, "Science", 8.8)
typeof(tup2)
# Tuple{String,Float64,String,Float64}

## variable assignment
a,b,c = ("Fast", 1, 5.2)
a
#"Fast"
 b
# 1
 c
#5.2
```

2.4 DATA STRUCTURES

2.4.1 Arrays

An array is a multidimensional grid that stores objects of any type. To improve performance, arrays should contain only one specific type, e.g., Int. Arrays do not require vectorizing for fast array computations. The array implementation used by Julia is written in Julia and relies on the compiler for performance. The compiler uses type inference to make optimized code for array indexing, which makes programs more readable and easier to maintain. Arrays are a subtype of the AbstractArray type. As such, they are laid out as a contiguous block of memory. This is not true of other members of the AbstractArray type, such as SparseMatrixCSC and SubArray.

The type and dimensions of an array can be specified using Array{T}(D), where T is any valid Julia type and D is the dimension of the array. The first entry in the tuple D is a singleton that specifies how the array values are initialized. Users can specify undef to create an uninitialized array, nothing to create arrays with no values, or missing to create arrays of missing values. Arrays with different types can be created with type Any.

```
## A vector of length 5 containing integers
a1 = Array{Int64}(undef, 5)

## A 2x2 matrix containing integers
a2 = Array{Int64}(undef, (2,2))
#2x2 Array{Int64,2}:
#493921239145  425201762420
#416611827821          104

## A 2x2 matrix containing Any type
a2 = Array{Any}(undef, (2,2))
#2x2 Array{Any,2}:
# #undef  #undef
# #undef  #undef
```

In Julia, [] can also be used to generate arrays. In fact, the Vector(), Matrix() and collect() functions can also be used.

```
## A three-element row "vector"
a4 = [1,2,3]

## A 1x3 column vector -- a two-dimensional array
a5 = [1 2 3]

## A 2x3 matrix, where ; is used to separate rows
a6 = [80 81 82 ; 90 91 92]
```

```
## Notice that the array a4 does not have a second dimension, i.e., it is
## neither a 1x3 vector nor a 3x1 vector. In other words, Julia makes a
## distinction between Array{T,1} and Array{T,2}.
a4
# 3-element Array{Int64,1}:
# 1
# 2
# 3

## Arrays containing elements of a specific type can be constructed like:
a7 = Float64[3.0 5.0 ; 1.1 3.5]

## Arrays can be explicitly created like this:
Vector(undef, 3)
# 3-element Array{Any,1}:
# #undef
# #undef
# #undef

Matrix(undef, 2,2)
#  2x2 Array{Any,2}:
#  #undef  #undef
#  #undef  #undef

## A 3-element Float array
a3 = collect(Float64, 3:-1:1)
# 3-element Array{Float64,1}:
# 3.0
# 2.0
# 1.0
```

Julia has many built-in functions that generate specific kinds of arrays. Here are some useful ones:

- `zeros(T, d1, ..)` is a d1-dimensional array of all zeros.

- `ones(T, d1, ..)` is a d1-dimensional array of all ones.

- `rand(T, d1, ..)`: if T is Float, a d1-dimensional array of random numbers between 0 and 1 is returned; if an array is specified as the first argument, d1 random elements from the array are returned.

- `randn(T, d1, ..)` is a d1-dimensional array of random numbers from the standard normal distribution with mean zero and standard deviation 1.

- `MatrixT(I, (n,n))` is the n×n identity matrix. The identity operator I is available in the `LinearAlgebra.jl` package.

- `fill!(A, x)` is the array A filled with value x.

Note that, in the above, d1 can be a tuple specifying multiple dimensions.

Arrays can easily be concatenated in Julia. There are two functions commonly used to concatenate arrays:

- vcat(A1, A2, ..) concatenates arrays vertically, i.e., stacks A1 on top of A2.

- hcat(A1, A2, ..) concatenates arrays horizontally, i.e., adds A2 to the right of A1.

Of course, concatenation requires that the relevant dimensions match.

The following code block illustrates some useful array functions as well as slicing. Slicing for arrays works similarly to slicing for strings.

```julia
## Create a 2x2 identity matrix
using LinearAlgebra
imat = Matrix{Int8}(I, (2,2))

## return random numbers between 0 and 1
rand(2)
#2-element Array{Float64,1}:
# 0.86398
# 0.491484

B = [80 81 82 ; 90 91 92]
# 2x3 Array{Int64,2}:
# 80   81   82
# 90   91   92

## return random elements of B
rand(B,2)
#2-element Array{Int64,1}:
# 80
# 91

## The number of elements in B
length(B)
# 6

## The dimensions of B
size(B)
# (2, 3)

## The number of dimensions of B
ndims(B)
# 2

## A new array with the same elements (data) as B but different dimensions
reshape(B, (3, 2))
# 3x2 Array{Int64,2}:
#  80   91
#  90   82
#  81   92

## A copy of B, where elements are recursively copied
```

```
B2 = deepcopy(B)

## When slicing, a slice is specified for each dimension
## The first two rows of the first column done two ways
B[1:2, ]
# 2-element Array{Int64,1}:
#  80
#  90

B[1:2,1]

## The first two rows of the second column
B[1:2,2]
# 2-element Array{Int64,1}:
#  81
#  91

## The first row
B[1,:]
# 3-element Array{Int64,1}:
#  80
#  81
#  82

## The third element
B[3]
#81

# Another way to build an array is using comprehensions
A1 = [sqrt(i) for i in [16,25,64]]
# 3-element Array{Float64,1}:
#  4.0
#  5.0
#  8.0

A2 = [i^2 for i in [1,2,3]]
# 3-element Array{Int64,1}:
#  1
#  4
#  9
```

From a couple of examples in the above code block, we can see that Julia counts array elements by column, i.e., the kth element of the $n \times m$ matrix $\mathbf{X}$ is the kth element of the nm-vector $\text{vec}(\mathbf{X})$. Array comprehensions, illustrated above, are another more sophisticated way of building arrays. They generate the items in the array with a function and a loop. These items are then collected into an array by the brackets [] that surround the loop and function.

2.4.2 Dictionaries

In Julia, dictionaries are defined as associative collections consisting of a key-value pair, i.e., the key is associated with a specific value. These key-value pairs have their own type in Julia, `Pairtypeof(key), typeof(value)` which creates a `Pair` object.

Alternatively, the => symbol can be used to separate the key and value to create the same `Pair` object. One use of `Pair` objects is in the instantiation of dictionaries. Dictionaries in Julia can be used analogously to lists in R. Dictionaries are created using the keyword `Dict` and types can be specified for both the key and the value. The keys are hashed and are always unique.

```
## Three dictionaries, D0 is empty, D1 and D2 are the same
D0 = Dict()
D1 = Dict(1 => "red", 2 => "white")
D2 = Dict{Integer, String}(1 => "red", 2 => "white")

## Dictionaries can be created using a loop
food = ["salmon", "maple syrup", "tourtiere"]

food_dict = Dict{Int, String}()

## keys are the foods index in the array
for (n, fd) in enumerate(food)
  food_dict[n] = fd
end

## Dictionaries can also be created using the generator syntax
wine = ["red", "white", "rose"]
wine_dict = Dict{Int,String}(i => wine[i] for i in 1:length(wine))
```

Values can be accessed using [] with a value of a dictionary key inserted between them or `get()`. The presence of a key can be checked using `haskey()` and a particular key can be accessed using `getkey()`. Keys can also be modified, as illustrated in the below code block. Here, we also demonstrate adding and deleting entries from a dictionary as well as various ways of manipulating keys and values. Note that the following code block builds on the previous one.

```
## Values can be accessed similarly to an array, but by key:
food_dict[1]

## The get() function can also be used; note that "unknown" is the
## value returned here if the key is not in the dictionary
get(food_dict, 1, "unknown")
get(food_dict, 7, "unknown")

## We can also check directly for the presence of a particular key
haskey(food_dict, 2)
haskey(food_dict, 9)

## The getkey() function can also be used; note that 999 is the
## value returned here if the key is not in the dictionary
getkey(food_dict, 1, 999)

## A new value can be associated with an existing key
```

```
food_dict
food_dict[1] = "lobster"

## Two common ways to add new entries:
food_dict[4] = "bannock"
get!(food_dict, 4, "bannock")

## The advantage of get!() is that is will not add the new entry if
## a value is already associated with the the key
get!(food_dict, 4, "toutiere")

## Just deleting entries by key is straightforward
delete!(food_dict,4)

## But we can also delete by key and return the value associated with
## the key; note that 999 is returned here if the key is not present
deleted_fd_value = pop!(food_dict,3, 999)

# Keys can be coerced into arrays
collect(keys(food_dict))

# Values can also be coerced into arrays
collect(values(food_dict))

# We can iterate over both keys and values
for (k, v) in food_dict
    println("food_dict: key: ", k, " value: ", v)
end

# We could also just loop over keys
for k in keys(food_dict)
    println("food_dict: key: ", k)
end

# Or could also just loop over values
for v in values(food_dict)
    println("food_dict: value: ", v)
end
```

2.5   CONTROL FLOW

2.5.1   Compound Expressions

In Julia, a compound expression is one expression that is used to sequentially evaluate a group of subexpressions. The value of the last subexpression is returned as the value of the expression. There are two ways to achieve this: begin blocks and chains.

```
## A begin block
b1 = begin
    c = 20
    d = 5
    c * d
end
println("b1: ", b1)
```

```
# 100

## A chain
b2 = (c = 20 ; d = 5 ; c * d)
println("b2: ", b2)
# 100
```

2.5.2   Conditional Evaluation

Conditional evaluation allows parts of a program to be evaluated, or not, based on the value of a Boolean expression, i.e., an expression that produces a true/false value. In Julia, conditional evaluation takes the form of an `if-elseif-else` construct, which is evaluated until the first Boolean expression evaluates to true or the `else` statement is reached. When a given Boolean expression evaluates to true, the associated block of code is executed. No other code blocks or condition expressions within the `if-elseif-else` construct are evaluated. An `if-elseif-else` construct returns the value of the last executed statement. Programmers can use as many `elseif` blocks as they wish, including none, i.e., an `if-else` construct. In Julia, `if`, `elseif` and `else` statements do not require parentheses; in fact, their use is discouraged.

```
# An if-else construct
k = 1
if k == 0
  "zero"
else
  "not zero"
end
# not zero

# An if-elseif-else construct
k = 11
if k % 3 == 0
  0
elseif k % 3 == 1
  1
else
  2
end
# 2
```

An alternative approach to conditional evaluation is via short-circuit evaluation. This construct has the form `a ? b : c`, where `a` is a Boolean expression, `b` is evaluated if `a` is true, and `c` is evaluated if `a` is false. Note that `? :` is called the "ternary operator", it associates from right to left, and it can be useful for short conditional statements. Ternary operators can be chained together to accommodate situations analogous to an `if-elseif-else` construct with one or more `ifelse` blocks.

```
# A short-circuit evaluation
b= 10; c = 20;
println("SCE: b < c: ", b < c ? "less than" : "not less than")

# A short-circuit evaluation with nesting
d = 10; f = 10;
println("SCE: chained d vs e: ",
  d < f ? "less than " :
  d > f ? "greater than" : "equal")

# Note that we do not use e in the above example because it is a literal
# in Julia (the exponential function); while it can be overwritten, it is
# best practice to avoid doing so.
e
# ERROR: UndefVarError: e not defined
using Base.MathConstants
e
# e = 2.7182818284590...
```

2.5.3 Loops

2.5.3.1 Basics

Two looping constructs exist in Julia: `for` loops and `while` loops. These loops can iterate over any container, such as a string or an array. The body of a loop ends with the **end** keyword. Variables referenced inside loops are typically in the local scope of the loop. When using variables defined outside the body of the loop, pre-append them with the **global** keyword inside the body of the loop. A `for` loop can operate over a range object representing a sequence of numbers, e.g., `1:5`, which it uses to get each index to loop through the range of values in the range, assigning each one to an indexing variable. The indexing variable only exists inside the loop. When looping over a container, `for` loops can access the elements of the container directly using the `in` operator. Rather than using simple nesting, nested `for` loops can be written as a single outer loop with multiple indexing variables forming a Cartesian product, e.g., if there are two indexing variables then, for each value of the first index, each value of the second index is evaluated.

```
str = "Julia"

## A for loop for a string, iterating by index
```

```
for i = 1:length(str)
  print(str[i])
end

## A for loop for a string, iterating by container element
for s in str
  print(s)
end

## A nested for loop
for i in str, j = 1:length(str)
  println((i, j))
end
# ('J', 1)
# ('J', 2)
# ..
# ('a', 4)
# ('a', 5)

## Another nested for loop
odd = [1,3,5]
even = [2,4,6]
for i in odd, j in even
  println("i*j: $(i*j)")
end
# i*j: 2
# i*j: 4
# ..
# i*j: 20
# i*j: 30
```

A while loop evaluates a conditional expression and, as long as it is true, the loop evaluates the code in the body of the loop. To ensure that the loop will end at some stage, an operation inside the loop has to falsify the conditional expression. Programmers must ensure that a while loop will falsify the conditional expression, otherwise the loop will become "infinite" and never finish executing.

```
## Example of an infinite while loop (nothing inside the loop can falsify
## the condition x<10)
n=0
x=1
while x<10:
  global n
  n=n+1
end

## A while loop to estimate the median using an MM algorithm
  using Distributions, Random
  Random.seed!(1234)

  iter = 0
  N = 100
  x = rand(Normal(2,1), N)
  psi = fill!(Vector{Float64}(undef,2), 1e9)
```

```
while(true)
    global iter, x, psi
    iter += 1
    if iter == 25
        println("Max iteration reached at iter=$iter")
        break
    end
    num, den = (0,0)
    ## elementwise operations in wgt
    wgt = (abs.(x .- psi[2])).^-1
    num = sum(wgt .* x)
    den = sum(wgt)
    psi = circshift(psi, 1)
    psi[2] = num / den

    dif = abs(psi[2] - psi[1])
    if dif < 0.001
        print("Converged at iteration $iter")
        break
    end
end

# gives an estimate of the median
median(x)
# 1.959

psi[2]
# 1.956
```

2.5.3.2 Loop termination

When writing loops, it is often advantageous to allow a loop to terminate early, before it has completed. In the case of a `while` loop, the loop would be broken before the test condition is falsified. When iterating over an iterable object with a `for` loop, it is stopped before the end of the object is reached. The `break` keyword can accomplish both tasks. The following code block has two loops, a `while` loop that calculates the square of the index variable and stops when the square is greater than 16. Note that without the `break` keyword, this is an infinite loop. The second loop does the same thing, but uses a `for` loop to do it. The `for` loop terminates before the end of the iterable range object is reached.

```
## break keyword

i = 0
while true
    global i
    sq = i^2
    println("i: $i --- sq: = $sq")
    if sq > 16
```

```
        break
    end
    i += 1
end

# i: 0 --- sq: = 0
# i: 1 --- sq: = 1
# i: 2 --- sq: = 4
# i: 3 --- sq: = 9
# i: 4 --- sq: = 16
# i: 5 --- sq: = 25

for i = 1:10
    sq = i^2
    println("i: $i --- sq: = $sq")
    if sq > 16
        break
    end
end
```

In some situations, it might be the case that a programmer wants to move from the current iteration of a loop immediately into the next iteration before the current one is finished. This can be accomplished using the `continue` keyword.

```
## continue keyword

for i in 1:5
    if i % 2 == 0
        continue
    end
    sq = i^2
    println("i: $i --- sq: $sq")
end

# i: 1 --- sq: 1
# i: 3 --- sq: 9
# i: 5 --- sq: 25
```

In real world scenarios, continue could be used multiple times in a loop and there could be more complex code after the `continue` keyword.

2.5.3.3 Exception handling

Exceptions are unexpected conditions that can occur in a program while it is carrying out its computations. The program may not be able to carry out the required computations or return a sensible value to its caller. Usually, exceptions terminate the function or program that generates it and prints some sort of diagnostic message to standard output. An example of this is given in the following code block, where we try and take the logarithm of a negative number and the `log()` function throws an exception.

```
## Generate an exception
log(-1)

# ERROR: DomainError with -1.0:
# log will only return a complex result if called with a complex argument.
#  Try log(Complex(x)).
# Stacktrace:
#  [1] throw_complex_domainerror(::Symbol, ::Float64) at ./math.jl:31
#  [2] log(::Float64) at ./special/log.jl:285
#  [3] log(::Int64) at ./special/log.jl:395
#  [4] top-level scope at none:0
```

In the above code block, the `log()` function threw a `DomainError` exception. Julia has a number of built-in exceptions that can be thrown and captured by a Julia program. Any exception can be explicitly thrown using the `throw()` function.

```
## throw()

for i in [1, 2, -1, 3]
  if i < 0
    throw(DomainError())
  else
    println("i: $(log(i))")
  end
end

# i: 0.0
# i: 0.6931471805599453
# ERROR: MethodError: no method matching DomainError()
# Closest candidates are:
#   DomainError(::Any) at boot.jl:256
#   DomainError(::Any, ::Any) at boot.jl:257
# Stacktrace:
#  [1] top-level scope at ./none:3

## error
for i in [1, 2, -1, 3]
  if i < 0
    error("i is a negative number")
  else
    println("i: $(log(i))")
  end
end

# i: 0.0
# i: 0.6931471805599453
# ERROR: i is a negative number
# Stacktrace:
#  [1] top-level scope at ./none:3
```

In the previous code block, we throw the DomainError() exception when the input to log() is negative. Note that DomainError() requires the brackets () to return an exception object. Without them, it is referring to the exception type. The error() function can be used in a similar way. It produces an object of type ErrorException that will immediately stop all execution of the Julia program.

If we want to test for an exception and handle it gracefully, we can use a try-catch statement to do this. These statements allow us to catch an exception, store it in a variable if required, and try an alternative way of processing the input that generated the exception.

```
## try/catch

for i in [1, 2, -1, "A"]
  try log(i)
  catch ex
    if isa(ex, DomainError)
      println("i: $i --- Domain Error")
      log(abs(i))
    else
      println("i: $i")
      println(ex)
      error("Not a DomainError")
    end
  end
end

# i: -1 --- Domain Error
# i: A
# MethodError(log, ("A",), 0x0000000000061f0)
# ERROR: Not a DomainError
# Stacktrace:
#  [1] top-level scope at ./none:10
```

In the previous code block, the exception is stored in the ex variable and when the error is not a DomainError(), its value is returned along with the ErrorException defined by the call to error(). Note that try-catch blocks can degrade the performance of code because of the overhead they require. For high-performance code, it is better to use standard conditional evaluation to handle known exceptions.

2.6 FUNCTIONS

A function is an object that takes argument values as a tuple and maps them to a return value. Functions are first-class objects in Julia. They can be:

- assigned to variables;

- called from these variables;

- passed as arguments to other functions; and

- returned as values from a function.

A first-class object is one that accommodates all operations other objects support. Operations typically supported by first-class objects in all programming languages are listed above. The basic syntax of a function is illustrated in the following code block.

```
function add(x,y)
    return(x+y)
end
```

In Julia, function names are all lowercase, without underscores, but can include Unicode characters. It is best practice to avoid abbreviations, e.g., `fibonacci()` is preferable to `fib()`. The body of the function is the part contained on the lines between the `function` and `end` keywords. Parenthesis syntax is used to call a function, e.g., `add(3, 5)` returns 8. Because functions are objects, they can be passed around like any value and, when passed, the parentheses are omitted.

```
addnew = add
addnew(3,5)
# 8
```

Functions may also be written in assignment form, in which case the body of the function must be a single expression. This can be a very useful approach for simple functions because it makes code much easier to read.

```
add2(x, y) = x+y
```

Argument passing is done by reference. Modifications to the input data structure (e.g., array) inside the function will be visible outside it. If function inputs are not to be modified by a function, a copy of the input(s) should be made inside the function before doing any modifications. Python and other dynamic languages handle their function arguments in a similar way.

```
## Argument passing
function f1!(x)
  x[1] = 9999
  return(x)
end

ia = Int64[0,1,2]
println("Array ia: ", ia)
# Array ia: [0, 1, 2]

f1!(ia)
println("Argument passing by reference: ", ia)
# Argument passing by reference: [9999, 1, 2]
```

By default, the last expression that is evaluated in the body of a function is its return value. However, when the function body contains one or more **return** keywords, it returns immediately when a **return** keyword is evaluated. The **return** keyword usually wraps an expression that provides a value when returned. When used with the control flow statements, the return keyword can be especially useful.

```
## A function with multiple options for return
function gt(g1, g2)
  if(g1 >g2)
     return("$g1 is largest")
  elseif(g1<g2)
     return("$g2  is largest")
  else
  return("$g1 and  $g2 are equal")
  end
end

gt(2,4)
# "4  is largest"
```

The majority of Julia operators are actually functions and can be called with parenthesized argument lists, just like other functions.

```
## These are equivalent
```

```
2*3
# 6
*(2,3)
# 6
```

Functions can also be created without a name, and such functions are called anonymous functions. Anonymous functions can be used as arguments for functions that take other functions as arguments.

```
## map() applies a function to each element of an array and returns a new
## array containing the resulting values

a = [1,2,3,1,2,1]
mu = mean(a)
sd = std(a)

## centers and scales a
b = map(x -> (x-mu)/sd, a)
```

Julia accommodates optional arguments by allowing function arguments to have default values, similar to R and many other languages. The value of an optional argument does not need to be specified in a function call.

```
## A function with an optional argument. This is a recursive function,
## i.e., a function that calls itself, for computing the sum of the first n
## elements of the Fibonacci sequence: 1, 1, 2, 3, 5, 8, 13, 21, 34, 55,...

function fibonacci(n=20)
  if (n<=1)
    return 1
  else
    return fibonacci(n-1)+fibonacci(n-2)
  end
end

## Sum the first 12 elements of the Fibonacci sequence
fibonacci(12)
# 233

## Because the optional argument defaults to 20, these are equivalent
fibonacci()
fibonacci(20)
```

Function arguments determine its behaviour. In general, the more arguments a function has, the more varied its behaviour will be. Keyword arguments are useful because they help manage function behaviour; specifically, they allow arguments to be specified by name and not just position in the function call. In the below code block, an MM algorithm is demonstrated. Note that we have already used an MM algorithm in Section 2.5.3.1, but now we construct an MM algorithm as a function. MM algorithms are blueprints for algorithms that either iteratively minimize a majorizing function or iteratively maximize a minorizing function — see Hunter and Lange (2000, 2004) for further details.

```
## A function with a keyword argument
## Arguments after the ; are keyword arguments
## The default values are evaluated from left-to-right.
## This allows keyword arguments to refer to previously defined keywords
## Keyword arguments can have explicit types

## estimate the median of a 1D array  using an MM algorithm
## for clarity (too many m's!) we use an _ in the function name
function mm_median(x, eps = 0.001; maxit = 25, iter::Int64=Int(floor(eps)))

    ## initalizations
    psi = fill!(Vector{Float64}(undef,2), 1e2)

    while(true)
      iter += 1
      if iter == maxit
          println("Max iteration reached at iter=$iter")
          break
      end
      num, den = (0,0)
      ## use map() to do elementwise operations in wgt
      wgt = map(d -> (abs(d - psi[2]))^(-1), x)
      num = sum(map(*, wgt, x))
      den = sum(wgt)
      psi = circshift(psi, 1)
      psi[2] = num / den

      dif = abs(psi[2] - psi[1])
      if dif < eps
          print("Converged at iteration $iter")
          break
      end
    end

    return(Dict(
      "psi_vec" => psi,
        "median" => psi[2]
    ))

end    .

## Run on simulated data
using Distributions, Random
Random.seed!(1234)

N = Int(1e3)
dat = rand(Normal(0,6), N)

## Function calls using different types of arguments
median(dat)
# 0.279
```

```
mm_median(dat, 1e-9)["median"]
# Max iteration reached at iter=25

mm_median(dat, maxit=50)["median"]
# Converged at iteration 26
# 0.296

mm_median(dat, 1e-9, maxit=100)["median"]
# Converged at iteration 36
# 0.288
```

Some general tips for using functions are as follows:

1. Write programs as a series of functions because: functions are testable and reusable, they have well-defined inputs and outputs, and code inside functions usually runs much faster because of how the Julia compiler works.

2. Functions should not operate on global variables.

3. Functions with ! at the end of their names modify their arguments instead of copying them. Julia library functions often come in two versions, distinguished by the !.

4. Pass a function directly, do not wrap it in an anonymous function.

5. When writing functions that take numbers as arguments, use the Int type when possible. When combined in numerical computations, they change the resulting Julia type of the result less frequently. This is known as type promotion.

6. If function arguments have specific types, the function call should force its arguments to be the required type.

The aforementioned tips are illustrated in the following code block.

```
## Tip3: Function with a ! in the name
a1 = [2,3,1,6,2,8]
sort!(a1)
a1
#6-element Array{Int64,1}:
# 1
# 2
# 2
# 3
# 6
# 8

## Tip 4
```

```
## Do not wrap abs() in an anonymous function
A = [1, -0.5, -2, 0.5]
map(x -> abs(x), A)

# Rather, do this
## abs() is not wrapped in an anonymous function
map(abs, A)

##Tip 5:  Type promotion
times1a(y) = *(y, 1)
times1b(y) = *(y, 1.0)
println("times1a(1/2): ", times1a(1/2))
println("times1a(2): ", times1a(2)) ## preserves type
println("times1a(2.0): ", times1a(2.0))
println("times1b(1/2): ", times1b(1/2))
println("times1b(2): ", times1b(2)) ## changes type
println("times1b(2.0): ", times1b(2.0))

## Tip6:  Function with typed arguments
times1c(y::Float64) = *(y, 1)
times1c(float(23))
```

## CHAPTER 3

Working with Data

THE PURPOSE of this chapter is to familiarize the reader with some of the basics of working with data in Julia. As would be expected, much of the focus of this chapter is on or around dataframes, including dataframe functions. Other topics covered include categorical data, input-output (IO), and the split-apply-combine strategy.

3.1 DATAFRAMES

A dataframe is a tabular representation of data, similar to a spread-sheet or a data matrix. As with a data matrix, the observations are rows and the variables are columns. Each row is a single (vector-valued) observation. For a single row, i.e., observation, each column represents a single realization of a variable. At this stage, it may be helpful to explicitly draw the analogy between a dataframe and the more formal notation often used in statistics and data science.

Suppose we observe n realizations $\mathbf{x}_1, \ldots, \mathbf{x}_n$ of p-dimensional random variables $\mathbf{X}_1, \ldots, \mathbf{X}_n$, where $\mathbf{X}_i = (X_{i1}, X_{i2}, \ldots, X_{ip})'$ for $i = 1, \ldots, n$. In matrix form, this can be written

$$
\mathscr{X} = (\mathbf{X}_1, \mathbf{X}_2, \ldots, \mathbf{X}_n)' = \begin{pmatrix} \mathbf{X}'_1 \\ \mathbf{X}'_2 \\ \vdots \\ \mathbf{X}'_n \end{pmatrix}
$$

$$
= \begin{pmatrix} X_{11} & X_{12} & \cdots & X_{1p} \\ X_{21} & X_{22} & \cdots & X_{2p} \\ \vdots & \vdots & \ddots & \vdots \\ X_{n1} & X_{n2} & \cdots & X_{np} \end{pmatrix}. \tag{3.1}
$$

Now, $\mathbf{X}_i$ is called a random vector and $\mathscr{X}$ is called an $n \times p$ random matrix. A realization of $\mathscr{X}$ can be considered a data matrix. For completeness, note that a matrix $\mathbf{A}$ with all entries constant is called a constant matrix.

Consider, for example, data on the weight and height of 500 people. Let $\mathbf{x}_i = (x_{i1}, x_{i2})'$ be the associated observation for the ith person, $i = 1, 2, \ldots, 500$, where x_{i1} represents their weight and x_{i2} represents their height. The associated data matrix is then

$$
\mathscr{X} = (\mathbf{x}_1, \mathbf{x}_2, \ldots, \mathbf{x}_{500})' = \begin{pmatrix} \mathbf{x}_1' \\ \mathbf{x}_2' \\ \vdots \\ \mathbf{x}_{500}' \end{pmatrix} = \begin{pmatrix} x_{11} & x_{12} \\ x_{21} & x_{22} \\ \vdots & \vdots \\ x_{500,1} & x_{500,2} \end{pmatrix}. \quad (3.2)
$$

A dataframe is a computer representation of a data matrix. In Julia, the `DataFrame` type is available through the `DataFrames.jl` package. There are several convenient features of a `DataFrame`, including:

- columns can be different Julia types;

- table cell entries can be missing;

- metadata can be associated with a `DataFrame`;

- columns can be names; and

- tables can be subsetted by row, column or both.

The columns of a `DataFrame` are most often integers, floats or strings, and they are specified by Julia symbols.

```
## Symbol versus String
fruit = "apple"

println("eval(:fruit): ", eval(:fruit))
# eval(:fruit): apple

println("""eval("apple"): """, eval("apple"))
# eval("apple"): apple
```

In Julia, a symbol is how a variable name is represented as data; on the other hand, a string represents itself. Note that `df[:symbol]` is how a column is accessed with a symbol; specifically, the data in the column represented by `symbol` contained in the `DataFrame` `df` is being accessed. In Julia, a `DataFrame` can be built all at once or in multiple phases.

```
## Some examples with DataFrames

using DataFrames, Distributions, StatsBase, Random

Random.seed!(825)

N = 50

## Create a sample dataframe
## Initially the DataFrame has N rows and 3 columns
df1 = DataFrame(
  x1 = rand(Normal(2,1), N),
  x2 = [sample(["High", "Medium", "Low"],
              pweights([0.25,0.45,0.30])) for i=1:N],
  x3 = rand(Pareto(2, 1), N)
  )

## Add a 4th column, y, which is dependent on x3 and the level of x2
df1[:y] = [df1[i,:x2] == "High" ? *(4, df1[i, :x3]) :
              df1[i,:x2] == "Medium" ? *(2, df1[i, :x3]) :
                *(0.5, df1[i, :x3])  for i=1:N]
```

A `DataFrame` can be sliced the same way a two-dimensional `Array` is sliced, i.e., via `df[row_range, column_range]`. These ranges can be specified in a number of ways:

- Using `Int` indices individually or as arrays, e.g., `1` or `[4,6,9]`.

- Using `:` to select indices in a dimension, e.g., `x:y` selects the range from `x` to `y` and `:` selects all indices in that dimension.

- Via arrays of Boolean values, where `true` selects the elements at that index.

Note that columns can be selected by their symbols, either individually or in an array `[:x1, :x2]`.

```
## Slicing DataFrames
println("df1[1:2, 3:4]: ",df1[1:2, 3:4])
println("\ndf1[1:2, [:y, :x1]]: ",df1[1:2, [:y, :x1]])

## Now, exclude columns x1 and x2
keep = setdiff(names(df1), [:x1, :x2])
println("\nColumns to keep: ", keep)
# Columns to keep: Symbol[:x3, :y]

println("df1[1:2, keep]: ",df1[1:2, keep])
```

In practical applications, missing data is common. In `DataFrames.jl`, the `Missing` type is used to represent missing values. In Julia, a singlton occurence of `Missing`, `missing` is used to represent missing data. Specifically, `missing` is used to represent the value of a measurement when a valid value could have been observed but was not. Note that `missing` in Julia is analogous to `NA` in R.

In the following code block, the array `v2` has type `Union{Float64, Missings.Missing}`. In Julia, `Union` types are an abstract type that contain objects of types included in its arguments. In this example, `v2` can contain values of `missing` or `Float64` numbers. Note that `missings()` can be used to generate arrays that will support missing values; specifically, it will generate vectors of type `Union` if another type is specified in the first argument of the function call. Also, `ismissing(x)` is used to test whether `x` is missing, where `x` is usually an element of a data structure, e.g., `ismissing(v2[1])`.

```
## Examples of vectors with missing values
v1 = missings(2)
println("v1: ", v1)
# v1: Missing[missing, missing]

v2 = missings(Float64, 1, 3)
v2[2] = pi
println("v2: ", v2)
# v2: Union{Missing, Float64}[missing 3.14159 missing]

## Test for missing
m1 = map(ismissing, v2)
println("m1: ", m1)
# m1: Bool[true false true]

println("Percent missing v2: ", *(mean([ismissing(i) for i in v2]), 100))
# Percent missing v2: 66.66666666666666
```

Note that most functions in Julia do not accept data of type `Missings.Missing` as input. Therefore, users are often required to remove them before they can use specific functions. Using `skipmissing()` returns an iterator that excludes the missing values and, when used in conjunction with `collect()`, gives an array of non-missing values. This approach can be used with functions that take non-missing values only.

```
## calculates the mean of the non-missing values
mean(skipmissing(v2))
# 3.141592653589793

## collects the non-missing values in an array
collect(skipmissing(v2))
# 1-element Array{Float64,1}:
#  3.14159
```

3.2 CATEGORICAL DATA

In Julia, categorical data is represented by arrays of type `CategoricalArray`, defined in the `CategoricalArrays.jl` package. Note that `CategoricalArray` arrays are analogous to factors in R. `CategoricalArray` arrays have a number of advantages over `String` arrays in a dataframe:

- They save memory by representing each unique value of the string array as an index.

- Each index corresponds to a level.

- After data cleaning, there are usually only a small number of levels.

`CategoricalArray` arrays support missing values. The type `CategoricalArray{Union{T, Missing}}` is used to represent missing values. When indexing/slicing arrays of this type, `missing` is returned when it is present at that location.

```
## Number of entries for the categorical arrays
Nca = 10

## Empty array
v3 = Array{Union{String, Missing}}(undef, Nca)

## Array has string and missing values
v3 = [isodd(i) ? sample(["High", "Low"], pweights([0.35,0.65])) :
        missing for i = 1:Nca]

## v3c is of type CategoricalArray{Union{Missing, String},1,UInt32}
v3c = categorical(v3)

## Levels should be ["High", "Low"]
println("1. levels(v3c): ", levels(v3c))
# 1. levels(v3c): ["High", "Low"]

## Reordered levels - does not change the data
levels!(v3c, ["Low", "High"])
println("2. levels(v3c):", levels(v3c))
# 2. levels(v3c): ["Low", "High"]

println("2. v3c: ", v3c)
# 2. v3c: Union{Missing, CategoricalString{UInt32}}
# ["High", missing, "Low", missing, "Low", missing, "High",
# missing, "Low", missing]
```

Here are several useful functions that can be used with `CategoricalArray` arrays:

- `levels()` returns the levels of the `CategoricalArray`.

- `levels!()` changes the order of the array's levels.

- `compress()` compresses the array saving memory.

- `decompress()` decompresses the compressed array.

- `categorical(ca)` converts the array `ca` into an array of type `CategoricalArray`.

- `droplevels!(ca)` drops levels no longer present in the array `ca`. This is useful when a dataframe has been subsetted and some levels are no longer present in the data.

- `recode(a, pairs)` recodes the levels of the array. New levels should be of the same type as the original ones.

- `recode!(new, orig, pairs)` recodes the levels in `orig` using the `pairs` and puts the `new` levels in `new`.

Note that ordered `CategoricalArray` arrays can be made and manipulated.

```
## An integer array with three values
v5 = [sample([0,1,2], pweights([0.2,0.6,0.2]))) for i=1:Nca]

## An empty string array
v5b = Array{String}(undef, Nca)

## Recode the integer array values and save them to v5b
recode!(v5b, v5, 0 => "Apple", 1 => "Orange", 2=> "Pear")
v5c = categorical(v5b)

print(typeof(v5c))
# CategoricalArray{String,1,UInt32,String,CategoricalString{UInt32},
# Union{}}

print(levels(v5c))
# ["Apple", "Orange", "Pear"]
```

3.3   INPUT/OUTPUT

The `CSV.jl` library has been developed to read and write delimited text files. The focus in what follows is on reading data into Julia with `CSV.read()`. However, as one would expect, `CSV.write()` has many of the same arguments as `CSV.read()` and it should be easy to use once one becomes familiar with `CSV.read()`. Useful `CSV.read()` parameters include:

- `fullpath` is a `String` representing the file path to a delimited text file.

- `Data.sink` is usually a `DataFrame` but can be any `Data.Sink` in the `DataStreams.jl` package, which is designed to efficiently transfer/stream "table-like" data. Examples of data sinks include arrays, dataframes, and databases (`SQlite.jl`, `ODBC.jl`, etc.).

- `delim` is a character representing how the fields in a file are delimited (`|` or `,`).

- `quotechar` is a character used to represent a quoted field that could contain the field or newline delimiter.

- `missingstring` is a string representing how the missing values in the data file are defined.

- `datarow` is an `Int` specifying at what line the data starts in the file.

- `header` is a `String` array of column names or an `Int` specifying the row in the file with the headers.

- `types` specifies the column types in an `Array` of type `DataType` or a dictionary with keys corresponding to the column name or number and the values corresponding to the columns' data types.

Before moving into an example using real data, we will illustrate how to change the user's working directory. R users will be familiar with the `setwd()` function, which sets the R session's working directory to a user-defined location. The following code block demonstrates how to set the user's working directory in Julia using the `cd()` function. We are using the function `homedir()` to prepend the path. Note that Windows users have to "Escape" their backslashes when specifying the path.

```
# Specify working directory
homedir()
# "/Users/paul"

cd("$(homedir())/Desktop")

pwd()
"/Users/paul/Desktop"

# On Windows
# cd("D:\\julia\\projects")
```

The following code block details how one could go about reading in and cleaning the `beer` data in Julia. We start by defining some Julia types to store the raw data. This was necessary as the raw data contained missing values in addition to valid entries. The column names that will be used by our dataframe are defined in an array. These same names are used as keys for the dictionary that defines the types for each column. The `CSV.jl` package is used to read the comma separated value (CSV) data into Julia and store it as a dataframe called `df_recipe_raw`. From the raw data, a cleaned version of the data is produced, with new columns for a binary outcome and dummy variables produced from the levels of the categorical variables.

```
using DataFrames, Query, CSV, JLD2, StatsBase, MLLabelUtils, Random
include("chp3_functions.jl")
Random.seed!(24908)

## Types for the files columns
IntOrMiss = Union{Int64,Missing}
FltOrMiss = Union{Float64,Missing}
StrOrMiss = Union{String,Missing}

## define variable names for each column
recipe_header = ["beer_id", "name", "url", "style", "style_id", "size",
    "og", "fg", "abv", "ibu", "color", "boil_size", "boil_time", "biol_grav",
    "efficiency", "mash_thick", "sugar_scale", "brew_method", "pitch_rate",
    "pri_temp", "prime_method", "prime_am"]

## dictionary of types for each column
recipe_types2 = Dict{String, Union}(
    "beer_id" => IntOrMiss,
    "name" => StrOrMiss,
    "url" => StrOrMiss,
    "style" => StrOrMiss,
    "style_id" => IntOrMiss,
    "size" => FltOrMiss,
    "og" => FltOrMiss,
    "fg" => FltOrMiss,
    "abv" => FltOrMiss,
    "ibu" => FltOrMiss,
    "color" => FltOrMiss,
    "boil_size" => FltOrMiss,
    "boil_time" => FltOrMiss,
    "biol_grav" => FltOrMiss,
    "efficiency" => FltOrMiss,
    "mash_thick" => FltOrMiss,
    "sugar_scale" => StrOrMiss,
    "brew_method" => StrOrMiss,
    "pitch_rate" => FltOrMiss,
    "pri_temp" => FltOrMiss,
    "prime_method" => StrOrMiss,
    "prime_am" => StrOrMiss
)

## read csv file
```

```
df_recipe_raw = CSV.read("recipeData.csv",
  DataFrame;
  delim = ',' ,
  quotechar = '"',
  missingstring = "N/A",
  datarow = 2,
  header = recipe_header,
  types = recipe_types2,
  allowmissing=:all
)

## Drop columns
delete!(df_recipe_raw, [:prime_method, :prime_am, :url])

#####
## Write the raw data dataframe
JLD2.@save "recipeRaw.jld2" df_recipe_raw

###########################
## Create  cleaned version

## Create a copy of the DF
df_recipe = deepcopy(df_recipe_raw)

## exclude missing styles
filter!(row -> !ismissing(row[:style]), df_recipe)

println("-- df_recipe: ",size(df_recipe))
# df_recipe: (73861, 19)

## Make beer categories
df_recipe[:y] = map(x ->
occursin(r"ALE"i, x) || occursin(r"IPA"i, x) || occursin(r"Porter"i, x)
  || occursin(r"stout"i, x) ? 0 :
occursin(r"lager"i, x) || occursin(r"pilsner"i, x) || occursin(r"bock"i, x)
  || occursin(r"okto"i, x) ? 1 : 99 ,
df_recipe[:style])

## remove styles that are not lagers or ales
filter!(row -> row[:y] != 99, df_recipe)

## remove extraneous columns
delete!(df_recipe, [:beer_id, :name, :style, :style_id])

## create dummy variables - one-hot-encoding
onehot_encoding!(df_recipe, "brew_method" , trace = true)
onehot_encoding!(df_recipe, "sugar_scale")

describe(df_recipe, stats=[:eltype, :nmissing])

delete!(df_recipe, [:brew_method,:sugar_scale])

JLD2.@save "recipe.jld2" df_recipe
```

The following code block illustrates many of the same steps used to read and clean the **food** data which is used for our regression examples in Chapters 5 and 7.

```
using DataFrames, Query, CSV, JLD2, StatsBase, MLLabelUtils, Random
include("chp3_functions.jl")
```

```julia
Random.seed!(24908)

## Types for the file columns
IntOrMiss = Union{Int64,Missing}
FltOrMiss = Union{Float64,Missing}
StrOrMiss = Union{String,Missing}

## define variable names for each column
food_header =
    ["gpa", "gender", "breakfast", "cal_ckn", "cal_day",
    "cal_scone", "coffee", "comfort_food", "comfort_food_reason",
    "comfoodr_code1", "cook", "comfoodr_code2", "cuisine", "diet_current",
    "diet_current_code", "drink", "eating_changes", "eating_changes_coded",
    "eating_changes_coded1", "eating_out", "employment", "ethnic_food",
    "exercise", "father_educ", "father_prof", "fav_cuisine",
    "fav_cuisine_code", "fav_food", "food_child", "fries", "fruit_day",
    "grade_level", "greek_food", "healthy_feeling", "healthy_meal",
    "ideal_diet", "ideal_diet_coded", "income", "indian_food",
    "italian_food", "life_reward", "marital_status", "meals_friend",
    "mom_educ", "mom_prof", "nut_check", "on_campus", "parents_cook",
    "pay_meal_out", "persian_food","self_perception_wgt", "soup", "sports",
    "thai_food", "tortilla_cal", "turkey_cal", "sports_type", "veggies_day",
    "vitamins", "waffle_cal", "wgt"]

## dictionary of types for each column
food_types = Dict{String, Union}(
    "gpa" => FltOrMiss,
    "gender" => IntOrMiss,
    "breakfast" => IntOrMiss,
    "cal_ckn" => IntOrMiss,
    "cal_day" => IntOrMiss,
    "cal_scone" => IntOrMiss,
    "coffee" => IntOrMiss,
    "comfort_food" => StrOrMiss,
    "comfort_food_reason" => StrOrMiss,
    "comfoodr_code1" => IntOrMiss,
    "cook" => IntOrMiss,
    "comfoodr_code2" => IntOrMiss,
    "cuisine" => IntOrMiss,
    "diet_current" => StrOrMiss,
    "diet_current_code" => IntOrMiss,
    "drink" => IntOrMiss,
    "eating_changes" => StrOrMiss,
    "eating_changes_coded" => IntOrMiss,
    "eating_changes_coded1" => IntOrMiss,
    "eating_out" => IntOrMiss,
    "employment" => IntOrMiss,
    "ethnic_food" => IntOrMiss,
    "exercise" => IntOrMiss,
    "father_educ" => IntOrMiss,
    "father_prof" => StrOrMiss,
    "fav_cuisine" => StrOrMiss,
    "fav_cuisine_code" => IntOrMiss,
    "fav_food" => IntOrMiss,
    "food_child" => StrOrMiss,
    "fries" => IntOrMiss,
    "fruit_day" => IntOrMiss,
    "grade_level" => IntOrMiss,
    "greek_food" => IntOrMiss,
    "healthy_feeling" => IntOrMiss,
    "healthy_meal" => StrOrMiss,
    "ideal_diet" => StrOrMiss,
```

```
  "ideal_diet_coded" => IntOrMiss,
  "income" => IntOrMiss,
  "indian_food" => IntOrMiss,
  "italian_food" => IntOrMiss,
  "life_reward" => IntOrMiss,
  "marital_status" => IntOrMiss,
  "meals_friend" => StrOrMiss,
  "mom_educ" => IntOrMiss,
  "mom_prof" => StrOrMiss,
  "nut_check" => IntOrMiss,
  "on_campus" => IntOrMiss,
  "parents_cook" => IntOrMiss,
  "pay_meal_out" => IntOrMiss,
  "persian_food" => IntOrMiss,
  "self_perception_wgt" => IntOrMiss,
  "soup" => IntOrMiss,
  "sports" => IntOrMiss,
  "thai_food" => IntOrMiss,
  "tortilla_cal" => IntOrMiss,
  "turkey_cal" => IntOrMiss,
  "sports_type" => StrOrMiss,
  "veggies_day" => IntOrMiss,
  "vitamins" => IntOrMiss,
  "waffle_cal" => IntOrMiss,
  "wgt" => FltOrMiss
)

## read csv file
df_food_raw = CSV.read("food_coded.csv",
  DataFrame;
  delim = ',' ,
  quotechar = '"',
  missingstrings = ["nan", "NA", "na", ""],
  datarow = 2,
  header = food_header,
  types = food_types,
  allowmissing=:all
)

## drop text fields which are not coded fields
delete!(df_food_raw, [:comfort_food, :comfort_food_reason, :comfoodr_code2,
  :diet_current, :eating_changes, :father_prof, :fav_cuisine, :food_child,
  :healthy_meal, :ideal_diet, :meals_friend, :mom_prof, :sports_type
])

## Change 1/2 coding to 0/1 coding
df_food_raw[:gender] = map(x -> x - 1, df_food_raw[:gender])
df_food_raw[:breakfast] = map(x -> x - 1, df_food_raw[:breakfast])
df_food_raw[:coffee] = map(x -> x - 1, df_food_raw[:coffee])
df_food_raw[:drink] = map(x -> x - 1, df_food_raw[:drink])
df_food_raw[:fries] = map(x -> x - 1, df_food_raw[:fries])
df_food_raw[:soup] = map(x -> x - 1, df_food_raw[:soup])
df_food_raw[:sports] = map(x -> x - 1, df_food_raw[:sports])
df_food_raw[:vitamins] = map(x -> x - 1, df_food_raw[:vitamins])

JLD2.@save "food_raw.jld2"  df_food_raw

##########################
## Create  cleaned version

## Create a copy of the DF
df_food = deepcopy(df_food_raw)
```

```
println("- df_food size: ", size(df_food))
# - df_food size: (125, 48)

## generate dummy variables
## used string array bc onehot_encoding!() takes a string
change2_dv = ["cal_ckn", "cal_day", "cal_scone", "comfoodr_code1",
    "cook", "cuisine", "diet_current_code", "eating_changes_coded",
    "eating_changes_coded1", "eating_out", "employment", "ethnic_food",
    "exercise", "father_educ", "fav_cuisine_code", "fav_food", "fruit_day",
    "grade_level", "greek_food", "healthy_feeling", "ideal_diet_coded",
    "income", "indian_food", "italian_food", "life_reward", "marital_status",
    "mom_educ", "nut_check", "on_campus", "parents_cook", "pay_meal_out",
    "persian_food", "self_perception_wgt", "thai_food", "tortilla_cal",
    "turkey_cal", "veggies_day", "waffle_cal"]

println("-- onehotencoding()")
for i in change2_dv
    println("i: ", i)
    onehot_encoding!(df_food, i)
    delete!(df_food, Symbol(i))
end

## remove NaNs
df_food[:gpa] =
    collect(FltOrMiss, map(x -> isnan(x)?missing:x, df_food[:gpa]))
df_food[:wgt] =
    collect(FltOrMiss, map(x -> isnan(x)?missing:x, df_food[:wgt]))

## remove missing gpa
filter!(row -> !ismissing(row[:gpa]), df_food)

println("--- df_food: ", size(df_food))
# --- df_food: (121, 214)

JLD2.@save "food.jld2" df_food
```

3.4   USEFUL DATAFRAME FUNCTIONS

There are several dataframe functions that have not been mentioned yet but that are quite useful:

- eltype() provides the types for each element in a DataFrame.

- head(df, n) displays the top n rows.

- tail(df, n) displays the bottom n rows.

- size(df) returns a tuple with the dimensions of the DataFrame.

- size(df, 1) returns the number of columns.

- size(df, 2) returns the number of rows.

- `describe(df)` returns statistical summary measures along with column types for each column in the `df`.

- `colwise(f, df)` applies function `f` to the columns in `df`.

- `delete!(df, col_symbol)` removes one or more columns, where columns are referenced by a symbol or an array of symbols, e.g., `:x1` or `[:x1, :x2]`.

- `rename!(df, :old_name => :new_name)` uses a `Pair` data structure to specify the existing name and its new name. The `Pair` data structure can be created dynamically:

```
rename!(df1, o => n for (o, n) = zip([:x1, :x2, :x3, :y],
    [:X1, :X2, :X3, :Y]))
```

- `filter(f, df)` filters the rows in dataframe `df` using the anonymous function `f` and returns a copy of `df` with the rows filtered by removing elements where `f` is false.

- `filter!(f, df)` updates the dataframe `df`; note that no copy is created.

```
## remove rows where the style column is missing.
filter!(row -> !ismissing(row[:style]), df_recipe)
```

- `push!(df, item)` adds one or more items `item` to the dataframe `df` that are not already in a dataframe.

- `append!(df1, df2)` adds dataframe `df2` to dataframe `df1`.

Several functions listed here have clear analogues in R. For example, the `describe()` function in Julia is similar to the `summary()` function in R. Similarly, the `size()` function in Julia is similar to the `dim()` function in R.

```
## using the dataframe previously defined

describe(df1[:X1])

# Summary Stats:
# Mean:           2.078711
```

```
# Minimum:          -0.229097
# 1st Quartile:     1.262696
# Median:           2.086254
# 3rd Quartile:     2.972752
# Maximum:          4.390025
# Length:           50
# Type:             Float64

## number of rows and columns of df_1
size(df1)
# (50, 4)
```

3.5 SPLIT-APPLY-COMBINE STRATEGY

Often data scientists need to extract summary statistics from the data in dataframes. The split-apply-combine (SAC) strategy is a convenient way to do this. This strategy for data analysis was outlined by Wickham (2011) and is implemented as the `plyr` package (Wickham, 2016) for R. The strategy involves partitioning the dataset into groups and administering some function to the data in each group and then recombining the results. Julia implements this strategy with one of:

- `by(df, cols, f)`,

- `groupby(df, cols, skipmissing = false)`, or

- `aggregate(df, cols, f)`.

The function `by(df, cols, f)` is used to apply function `f` to the dataframe `df`, partitioning by the column(s) `cols` in the dataframe. The function `by()` takes the following arguments:

- `df` is the dataframe being analyzed.

- `cols` is the columns making up the groupings.

- `f` is the function being applied to the grouped data.

The function `by(df, cols, f)` returns a `DataFrame` of the results.

```
## a count of the levels of X2
## the counts are in column x1 of the dataframe returned from by()
by(df1, :X2, nrow )

# 3x2 DataFrame
# | Row | X2     | x1    |
# |     | String | Int64 |
# +-----+--------+-------+
```

```
# | 1    | Medium | 28    |
# | 2    | Low    | 11    |
# | 3    | High   | 11    |

## median of X3 by the levels of X2
by(df1, :X2, df -> DataFrame(Median = median(df[:X3])))

# 3x2 DataFrame
# | Row | X2     | Median  |
# |     | String | Float64 |
# +-----+--------+---------+
# | 1   | Medium | 1.21086 |
# | 2   | Low    | 1.19345 |
# | 3   | High   | 1.82011 |
```

The function groupby(df, cols, skipmissing = false) splits a dataframe df into sub-dataframes by rows and takes the following arguments:

- df is the DataFrame to be split.

- cols is the columns by which to split up the dataframe.

- skipmissing determines if rows in cols should be skipped if they contain missing entries and returns a grouped DataFrame object that can be iterated over, returning sub-dataframes at each iteration.

The function groupby(df, cols, skipmissing = false) returns a grouped DataFrame object, each sub-dataframe in this object is one group, i.e., a DataFrame, and the groups are accessible via iteration.

```
## print the summary stats for x3 in each partition
for part in groupby(df1, :X2, sort=true)
  println(unique(part[:X2]))
  println(summarystats(part[:X3]))
end

# ["High"]
# Summary Stats:
# Mean:           2.004051
# Minimum:        1.011101
# 1st Quartile:   1.361863
# Median:         1.820108
# 3rd Quartile:   2.383068
# Maximum:        4.116220
#
# ["Low"]
# ...
```

The function aggregate(df, cols, f) splits a dataframe df into sub-dataframes by rows and takes the following arguments:

- `df` is the dataframe being analyzed.

- `cols` are the columns that make up the groupings.

- `f` is the function to apply to the remaining data. Multiple functions can be specified as an array, e.g., `[sum, mean]`.

The function `aggregate(df, cols, f)` returns a `DataFrame`.

```
## keep the grouping variable X2  and Y
keep2 = setdiff(names(df1), [:X1, :X3])

## agg_res has the summary statistics by levels of X2
## MAD = median absolute deviation
agg_res = aggregate(df1[keep2], [:X2],[length, mean, std, median, mad])
rename!(agg_res, :Y_length => :Y_n)
agg_res

# 3x6 DataFrame
# | Row | X2     | Y_n   | Y_mean   | Y_std    | Y_median | Y_mad    |
# |     | String | Int64 | Float64  | Float64  | Float64  | Float64  |
# +-----+--------+-------+----------+----------+----------+----------+
# | 1   | Medium | 28    | 3.22175  | 1.70015  | 2.42171  | 0.473198 |
# | 2   | Low    | 11    | 0.692063 | 0.208037 | 0.596726 | 0.109055 |
# | 3   | High   | 11    | 8.01621  | 3.65864  | 7.28043  | 3.39415  |
```

Often dataframes need to be sorted. This can be accomplished with the `sort!()` function. The `!` in the function names indicates it will sort the object in place and not make a copy of it. When sorting dataframes, users will most often want to select the columns to sort by and the direction to sort them (i.e., ascending or descending). To accomplish this, the `sort!()` function takes the following arguments:

- `df` is the dataframe being sorted.

- `cols` is the dataframe columns to sort. These should be column symbols, either alone or in an array.

- `rev` is a Boolean value indicating whether the column should be sorted in descending order or not.

```
## sorting dataframes
sort!(df1, [:X2, :Y], rev = (true, false))

# 50x4 DataFrame
# | Row | X1      | X2     | X3      | Y       |
# |     | Float64 | String | Float64 | Float64 |
# +-----+---------+--------+---------+---------+
# | 1   | 1.45373 | Medium | 1.00982 | 2.01964 |
```

```
#  |  2  |  3.11033  |  Medium  |  1.01574  |  2.03148  |
#  |  3  |  2.12326  |  Medium  |  1.01782  |  2.03563  |
#  :
#  |  45  |  2.31324  |  High   |  1.82011  |  7.28043  |
#  :
#  |  50  |  2.33929  |  High   |  4.11622  |  16.4649  |
```

3.6   QUERY.JL

Query.jl is a Julia package used for querying Julia data sources. These data sources include the ones we have mentioned, such as dataframes and data streams such as CSV. They can also include databases via SQLite and ODBS, and time series data through the TimeSeries.jl framework. Query.jl can interact with any iterable data source supported through the IterableTables.jl package. It has many features that will be familiar to users of the dplyr package (Wickham et al., 2017) in R.

At the time of writing, Query.jl is heavily influenced by the query expression portion of the C# Language-INtegrated Query (LINQ) framework (Box and Hejlsberg, 2007). LINQ is a component of the .NET framework that allows the C# language to natively query data sources in the form of query expressions. These expressions are comparable to SQL statements in a traditional database and allow filtering, ordering and grouping operations on data sources with minimal code. The LINQ framework allows C# to query multiple data sources using the same query language. Query.jl gives data scientists this capability in Julia, greatly simplifying their work.

The following code block shows the basic structure of a Query.jl query statement. The @from statement is provided by the package which specifies how the query will iterate over the data source. This is done in the same way a for loop iterates over a data source, using a range variable, often i, the in operator and the data source name. There can be numerous query statements where <statements> is written in the below code block, each one would be separated by a new line. The result of the query is stored in the x_obj object which can be a number of different data sinks, including dataframes and CSV files.

```
## Pseudo code for a generic Query.jl statement
## the query statements in <statements> are separated by \n
x_obj  = @from <range_var> in <data_source> begin
    <statements>
end
```

We will start our overview of `Query.jl` features with a simple example using the `beer` data. The following code block shows how to filter rows with the `@where` statement, select columns with the `@select` statement, and return the result as a `DataFrame` object with the `@collect` statement.

```
using Query
## select lagers (y ==1) and pri_temp >20
x_obj = @from i in df_recipe begin
    @where i.y == 1 && i.pri_temp > 20
    @select {i.y, i.pri_temp, i.color}
    @collect DataFrame
end

typeof(x_obj)
# DataFrame

names(x_obj)
# Symbol[3]
# :y
# :pri_temp
# :color

size(x_obj)
# (333, 3)
```

Notice that the `@where` and `@select` statements use the iteration variable `i` to reference the columns in the dataframe. The dataframe `df_recipe` is the data source for the query. The `@collect` macro returns the result of the query as an object of a given format. The formats available are called data sinks and include arrays, dataframes, dictionaries or any data stream available in `DataStreams.jl`, amongst others.

The `@where` statement is used to filter the rows of the data source. It is analogous to the `filter!` function described in Section 3.3. The expression following `@where` can be any arbitrary Julia expression that evaluates to a Boolian value.

The `@select` statement can use named tuples, which are integrated into the base language. They are tuples that allow their elements to be accessed by an index or a symbol. A symbol is assigned to an element of the tuple, and `@select` uses these symbols to construct column names in the data sink being used. Named tuples are illustrated in the following code block. The second element is accessed via its symbol, unlike a traditional tuple which would only allow access via its index (e.g., `named_tup[2]`). Named tuples can be defined in a `@select` statement in two ways, using the traditional (name = value) syntax or the custom Query.jl syntax {name = value}.

```
## named tuple - reference property by its symbol
named_tup = (x = 45, y =90)

typeof(named_tup)
# NamedTuple{(:x, :y),Tuple{Int64,Int64}}

named_tup[:y]
# 90
```

The next code block shows how to query a dataframe and return different arrays. The first query filters out the rows of the dataframe, uses the get() function to extract the values of the colour column and returns them as an array. Arrays are returned when no data sink is specified in the @collect statement. The second query selects two columns of the dataframe and returns an array. This array is an array of named tuples. To create a two-dimensional array of floating point values, additional processing is required. The loop uses the arrays' indices and the symbols of the named tuple to populate the array of floats.

```
## Returns an array of Float64 values
a1_obj = @from i in df_recipe begin
    @where i.y == 1 && i.color <= 5.0
    @select get(i.color) #Float64[1916]
    @collect
end

a1_obj
# 1898-element Array{Float64,1}:
#  3.3
#  2.83
#  2.1

## Returns a Named Tuple array. Each row is a NT with col and ibu values
a2_obj = @from i in df_recipe begin
    @where i.y == 1 && i.color <= 5.0
    @select {col = i.color, ibu = i.ibu}
    @collect
end

a2_obj
# 1898-element Array{NamedTuple{(:col, :ibu),Tuple{DataValues.DataValue{
#  Float64},DataValues.DataValue{Float64}}},1}:
#  (col = DataValue{Float64}(3.3), ibu = DataValue{Float64}(24.28))
#  (col = DataValue{Float64}(2.83), ibu = DataValue{Float64}(29.37))

## Additional processing to return an Array of floats
N = size(a2_obj)[1]
a2_array =zeros(N, 2)
for (i,v) in enumerate(a2_obj)
```

```
    a2_array[i, 1] = get(a2_obj[i][:col],0)
    a2_array[i, 2] = get(a2_obj[i][:ibu],0)
end

a2_array
# 1898x2 Array{Float64,2}:
#  3.3   24.28
#  2.83  29.37
```

Another common scenario is querying a data structure and returning a dictionary. This can be accomplished in Query.jl by specifying the Dict type in the @collect statement. In addition to this, the @select statement should include a pair expression, the first variable being the dictionary's key, the second being its value.

```
## data sink is dictionary
## select statement creates a Pair
dict_obj = @from i in df_recipe begin
    @where i.y == 1 && i.color <= 5.0
    @select i.id => get(i.color)
    @collect Dict
end

typeof(dict_obj)
# Dict{String,Float64}

dict_obj
# Dict String -> Float64 with 1898 entries
# "id16560" -> 3.53
# "id31806" -> 4.91
# "id32061" -> 3.66
```

In Query.jl, the @let statement can be used to create new variables in a query, known as range variables. The @let statement is used to apply a Julia expression to the elements of the data source and writes them to a data sink. The following code block details how to do this in the case where the objective is to mean centre and scale each column to have a mean of 0 and a standard deviation of 1. The @let statement can have difficulty with type conversion, so we define a constant dictionary with specified types. We store the column means and standard deviations here. The @let statement uses the dictionary values to do the centering and scaling.

```
using Statistics, StatsBase

## all missing values are skipped in the calculations
## Use a typed const to ensure type inference occurs correctly
```

```
const cs_dict = Dict{String, Float64}()
push!(cs_dict, "m_color" => mean(skipmissing(df_recipe[:color])))
push!(cs_dict, "m_ibu" => mean(skipmissing(df_recipe[:ibu])))
push!(cs_dict, "sd_color" => std(skipmissing(df_recipe[:color])))
push!(cs_dict, "sd_ibu" => std(skipmissing(df_recipe[:ibu])))

## mean center and scale a column and return as array
s1_obj = @from i in df_recipe begin
    @let ibu_cs = (i.ibu - cs_dict["m_ibu"]) / cs_dict["sd_ibu"]
    @select get(ibu_cs, missing)
    @collect
end
s1_obj
# 50562-element Array{Union{Missing, Float64},1}:
# -0.8151417763351124
# 0.1281156968892236
# 0.01995852988729847
# :
# -0.6337461084073555
# 0.07913886654872931

mean(skipmissing(s1_obj))
# 1.1198281324600945e-14
std(skipmissing(s1_obj))
# 1.00000000000000000000

## use named tuples
s2_obj = @from i in df_recipe begin
    @let ibu_cs = (i.ibu - cs_dict["m_ibu"]) / cs_dict["sd_ibu"]
    @let color_cs = (i.color - cs_dict["m_color"]) / cs_dict["sd_color"]
    @select {id = i.id, ibu = ibu_cs, color = color_cs}
    @collect DataFrame
end

s2_obj
# 50562x3 DataFrame
# | Row   | id       | ibu        | color      |
# |       | String   | Float64    | Float64    |
# +-------+----------+------------+------------+
# | 1     | id1      | -0.815142  | -0.773402  |
# | 2     | id2      | 0.128116   | -0.453797  |
# | 3     | id3      | 0.0199585  | -0.490763  |
# :
# | 50560 | id50560  | 0.127209   | -0.536971  |
# | 50561 | id50561  | -0.633746  | -0.0579493 |
# | 50562 | id50562  | 0.0791389  | -0.479211  |

mean(skipmissing(s2_obj[:color]))
# -5.692740670427803e-15
std(skipmissing(s2_obj[:color]))
# 0.9999999999999948
```

Sorting is a common task when working with data. `Query.jl` provides the `@orderby` statement to do sorting. It sorts the data source by one or more variables and the default sort order is ascending. If multiple variables are specified, they are separated by commas and the data source is sorted first by the initial variable in the specification. The `descending()` function can be used to change each variable's sorting order. Sorting is detailed in the following code block.

```
## sort at dataframe by 2 variables, one in ascending order
sort_obj = @from i in df_recipe begin
    @orderby i.y, descending(i.color)
    @select {i.y, i.color, i.ibu}
    @collect DataFrame
end

sort_obj

# 50562x3 DataFrame
# | Row   | y     | color    | ibu     |
# |       | Int64 | Float64  | Float64 |
# +-------+-------+----------+---------+
# | 1     | 0     | missing  | missing |
# | 2     | 0     | missing  | missing |
# :
# | 8     | 0     | 186.0    | missing |
# :
# | 50561 | 1     | 0.11     | 91.14   |
# | 50562 | 1     | 0.03     | missing |
```

When working with multiple datasets, combining them is often necessary before the data can be analyzed. The @join statement is used to do this and implements many of the traditional database joins. We will illustrate an inner join in the next code block. Left outer and group joins are also available. The @join statement creates a new range variable j for the second data source and uses id as the key. This key is compared to a key from the first data source y and matches are selected. Inner joins return all the rows that share the specified key values in both data sources, which in this case is all the rows in df_recipe.

```
## dataframe of beer labels
beer_style = DataFrame(id = 0:1, beername = ["Ale","Lager"])

## inner join
j1_obj = @from i in df_recipe begin
    @join j in beer_style on i.y equals j.id
    @select {i.y, j.beername}
    @collect DataFrame
end

j1_obj
# 50562x2 DataFrame
# | Row   | y     | beername |
# |       | Int64 | String   |
# +-------+-------+----------+
# | 1     | 0     | Ale      |
# | 2     | 0     | Ale      |
# | 3     | 0     | Ale      |
# :
# | 50562 | 0     | Ale      |
```

When processing data, it is often necessary to group the data into categories and calculate aggregate summaries for these groups. Query.jl facilitates this with the @group statement. It groups data from the data source by levels of the specified columns into a new range variable. This range variable is used to aggregate the data. In the following code block, we group the df_recipe dataframe by the beer categories y. The new range variable is called grp and is used in the @select statement to specify which data are used. The @where statement filters out the missing values in the IBU variable. In the query, missing values are represented as instances of the Query.jl data type DataValue. Consequently, isna() from the DataValues.jl package is used to filter them out and not ismissing(). The data for each group is aggregated by its mean and trimmed mean values.

```
using DataValues

## group by beer type and summarise ibu
## filter out missing values
g1_obj = @from i in df_recipe begin
    @where !isna(i.ibu)
    @group i by i.y into grp
    @select {Group = key(grp), Mean_IBU = mean(grp.ibu),
            TrimM_IBU = mean(trim(grp.ibu, prop=0.2))}
    @collect DataFrame
end

# 2x3 DataFrame
# | Row | Group | Mean_IBU | TrimM_IBU |
# |     | Int64 | Float64  | Float64   |
# +-----+-------+----------+-----------+
# | 1   | 0     | 55.6269  | 47.8872   |
# | 2   | 1     | 33.9551  | 29.1036   |
```

## CHAPTER 4

Visualizing Data

D ATA VISUALIZATION is a crucially important part of data science. This chapter outlines how several different types of data visualizations may be carried out in Julia. This includes very well-known approaches such as histograms and boxplots as well as perhaps lesser-known, but very useful, approaches such as violin plots and hexbin plots. Ancillary topics, such as saving plots in Julia, are also discussed.

4.1 GADFLY.JL

Julia has a plethora of high-quality plotting options. For those familiar with R, GadFly.jl is a good option. GadFly.jl is an implementation of the Grammar of Graphics (GoG; Wilkinson, 2005) written entirely in Julia. It will be very familiar to anyone who regularly uses the popular ggplot2 package (Wickham, 2009) in R.

The GoG framework provides a formal definition for creating static visualizations. It breaks visualizations into component parts similar to sentences in a human language. Sentences can be decomposed into nouns, adjectives and verbs to form a grammar for that sentence. These three components can be combined in specific ways to produce sentences with very different meanings. For example, if we take the nouns from

"fast Julia passed slow R"

and combine them with new adjectives and verbs we can get very different meanings. Another example along similar lines is

"popular R overshadows Julia".

GoG allows us to do the same thing with graphics, by taking a set of nouns (data) and combining them with adjectives and verbs (scales, geometries, etc.) to create both well-known visualizations and custom visualizations to meet specific needs. All plots in the GoG framework are built on data, aesthetics (scales to map the data on, e.g., $\log_{10}$) and geometries (points, lines, etc.). In addition to these three components, users can add facets (row and column subplots), statistics, coordinates (the plotting space used) and themes (non-data related elements) to their plots to fully customize them. This allows data scientists to quickly probe and explore their data and plan their visualizations the same way they would plan their data analyses.

In this chapter, we will cover exploratory graphics through the lens of the capabilities of GadFly.jl. Exploratory graphics is a vast subject with many tomes devoted to it. It is most often used to explore preexisting ideas about data and find interesting patterns that can help direct modelling efforts. Exploratory graphics typically do not make any parametric assumptions about the data being examined. At best, graphics can essentially reveal the answer to a question, e.g., a side-by-side boxplot can be very revealing when comparing two means (as one might for an independent groups t-test) and a quantile-quantile plot, or QQ-plot, can essentially address a question about normality. Even when graphics do not essentially answer the question, they help the analyst gain an understanding of the data under consideration. They can also help discover, or at least point to, underlying structure in data.

Gadfly.jl has a simple plotting interface and prefers that its data is consumed as Julia dataframes, although other formats are possible. The interface is accessed through the plot() function detailed in the following code block. Note that the specification of Gadfly.plot() in this code block ensures that the plot() function from Gadfly.jl is used rather than another plot() function.

```
GadFly.plot(data::DataFrame, mapping::Dict, elements::Element)
```

The mapping components are aesthetics that map the columns of the dataframe to the geometry of the plot. The elements are the adjectives and verbs used to display the data. They include:

- Geometries, which take aesthetics as input and use the data bound to them to depict the plot.

- Guides, which draw the axis, ticks, labels and keys outside the plotting frame.

- Statistics, which take aesthetics as input, perform an operation on them and return the transformed aesthetic.

- Coordinates, which map data to the 2D visualization space.

- Scales, which map an aesthetic a transformation of an aesthetic back on to itself.

Plots can be saved as Julia objects, to be used by subsequent function calls. When doing this, the plot() function call can be postpended with a semicolon to keep the plot from rendering. This is illustrated in the following code block.

```
## Plot, referenced with p1, and not rendered to STDOUT
p1 = GadFly.plot(data::DataFrame, mapping::Dict, elements::Element);
```

We will illustrate all of our exploratory plots using the data introduced in Chapter 3. Simple plots or plots designed for small to moderately sized datasets will use the simulated data from Chapter 3. Plotting methods suitable for large datasets with greater than 10,000 rows will be illustrated using the beer data.

4.2 VISUALIZING UNIVARIATE DATA

Visualizations at their most basic level display one set of numbers, i.e., univariate data. Dot charts and bar charts do this by displaying a point or a bar whose position or height coincide to the number. GadFly.jl code to generate both types of plot is given in the following code block. We start by using the by() function to create a new dataframe and the result df_bc is used to store the count of each level of x2. The call to plot() to generate the bar plot in Figure 4.1 uses the bar Geom to draw the bars and the ylabel Guide to change the plot's y-axis label. The dot plot function call differs in two meaningful ways. First, it uses the point Geom to draw the points and the cartesian Coord to put the categories on y-axis rather than the x-axis. Note that the Geom, Guide, and Coord keywords are a shorthand for components of the GoG framework as implemented in Gadfly.jl. In either graph, the summary measure need not be the category's count, it could just as easily be another single-number summary, such as a variable's mean value for the levels of x2.

```
using Gadfly, Cairo

## adds the dark theme to the top of themes stack
Gadfly.push_theme(:dark)

## create a df of means of each level of x2.
df_bc = by(df_1, :x2, nrow)
rename!(df_bc, :x1 => :count)

## Geom.bar to draw the bars
## Guide.ylabel to rename the Y label
p_bar = plot(df_bc, x=:x2, y=:count, Guide.ylabel("Count"), Geom.bar,
    style(bar_spacing=1mm))

## Dot plot
## same data different Geom
p_dot = plot(df_bc, y=:x2, x=:count, Guide.xlabel("Count"), Geom.point,
    Coord.cartesian(yflip = true))
```

The bar and dot plots that result from this code block are shown in Figures 4.1 and 4.2, respectively.

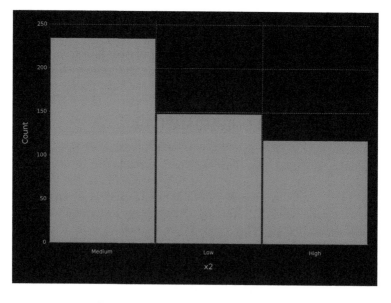

Figure 4.1   Bar plot of the counts for each level of x2.

Histograms are another common plot method for summarizing univariate data. A histogram is a bar chart where each bar represents a range of values of the continuous measurement. The height of the bar is determined by the number of values in the range. Because all the bars are the same width, the height of the bar is

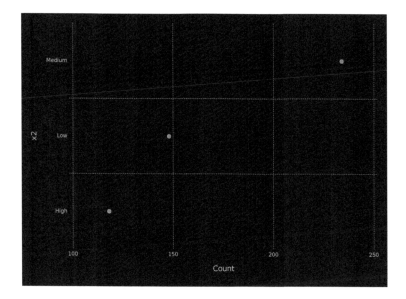

Figure 4.2   Dot plot of the counts for each level of x2.

equivalent to the proportion of observations in the range of values. The code to generate a histogram is very similar to the preceding calls to `plot()` except we are using a continuous variable for the x-axis aesthetic and the `histogram Geom` to render the plot.

```
## histogram
p_hist = plot(df_1, x=:x1, Guide.ylabel("Count"),
  Geom.histogram(bincount=10), style(bar_spacing=1mm))
```

The `histogram Geom` has a number of options; in the previous code block, here we use `bincount` to specify the number of bins to draw. The resulting histogram is shown in Figure 4.3.

Histograms are a coarse way of approximating a variable's probability density function, or density. Kernel density estimates (Ruppert et al., 2003) are an alternative method of visualizing the density of a variable. They use a non-negative kernel function, which integrates to one, and a bandwidth parameter to approximate the density. The smaller the bandwidth, the more closely the function will fit the data. `Gadfly.jl` uses a normal density function as its kernel function. The following code block illustrates the creation

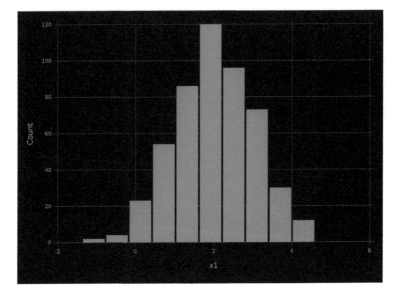

Figure 4.3 Histogram of x1 from the simulated data.

of a density plot. The only difference between it and the code used to generate the histogram is the `density` Geom.

```
## kernel density estimate
p_den = plot(df_1, x=:x1, Guide.ylabel("Density"),
    Geom.density(bandwidth=0.25), Guide.title("Bandwidth: 0.25"))

p_den2 = plot(df_1, x=:x1, Guide.ylabel("Density"),
    Geom.density(bandwidth=0.05), Guide.title("Bandwidth: 0.05"))

p_den3 = plot(df_1, x=:x1, Guide.ylabel("Density"),
    Geom.density(bandwidth=0.5), Guide.title("Bandwidth: 0.5"))
```

Figures 4.4, 4.5, and 4.6 illustrate different settings for the `bandwidth` using the bandwidth argument in `Geom`.

4.3 DISTRIBUTIONS

Data scientists often want to investigate the distributions within data. There are many options for doing this. The first and most well-known is the boxplot which gives a quick visual display of numeric data and is a good alternative to histograms. It was popularized by Tukey (1977) and consists of a rectangular box with lines or whiskers extending from the top and bottom. The box gives an idea

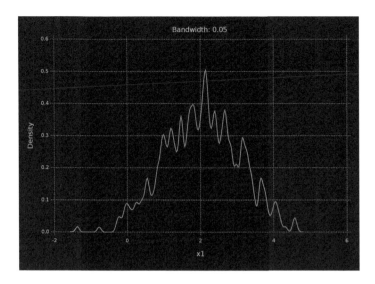

Figure 4.4 Kernel density estimate of x1 from the simulated data with bandwidth = 0.05.

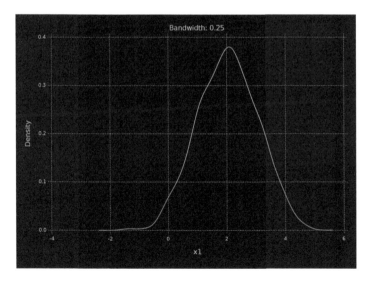

Figure 4.5 Kernel density estimate of x1 from the simulated data with bandwidth = 0.25.

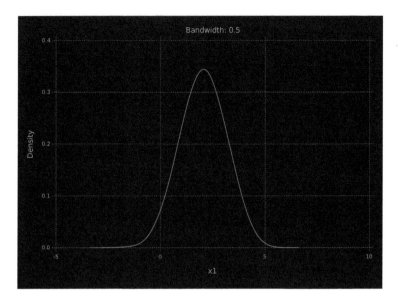

Figure 4.6 Kernel density estimate of x1 from the simulated data with bandwidth = 0.5.

of the location and spread on the central portion of the data. The box extends across the inter quartile range (IQR), with the middle line indicating the median value of the data. The whiskers extend 1.5 times the IQR above and below the box. Outlier values are indicated as points beyond the range covered by the whiskers. If the data are roughly normally distributed, approximately 99% will fall between the whiskers. Boxplots excel at comparing distributions between two or more categories and do not make any assumptions about the underlying distribution of the data. The following code block creates a boxplot from the beer data. It maps the x aesthetic to the beer categories and the y aesthetic to a quantitative measure of the beer colour and the boxplot Geom is used to depict the plot.

```
## boxplot
p_bp = plot(df_beer1, x=:c3, y =:color, Geom.boxplot(),
  Guide.xlabel("Type of Beer"), Guide.ylabel("Color"))

p_bp2 = plot(df_beer, x=:c3, y =:color, Geom.boxplot(),
  Guide.xlabel("Type of Beer"), Guide.ylabel("Color"))

p_bp3 = plot(df_beer, x=:c6, y =:color, Geom.boxplot(),
  Guide.xlabel("Type of Beer"), Guide.ylabel("Color"))
```

The resulting boxplots are shown in Figures 4.7–4.9. From these plots, many features of the beer data are apparent at a glance. For example, the boxplot in Figure 4.7 indicates that ales have substantially more variation in their colour values than lagers, and both types have outliers in the right tails of their distribution. From Figure 4.9, it is clear that porters and stouts are much darker than the other types of beer. Several other points are immediately apparent, e.g., with only a few exceptions, pilsners have relatively little variation in their colour and are very light, and while IPAs tend to be lighter in colour, their colours span the majority of the (beer) colour spectrum. It should be noted that, in Figures 4.7 and 4.8, stouts and IPAs were members of the ale group and pilsners were part of the lager group. Details on the characteristics of different beers are available from many sources, including Briggs et al. (2004).

Violin plots (Hintze and Nelson, 1998) are a modern alternative to boxplots. They look like water droplets, where the sides of the drop are made up of kernel density traces. Similar to boxplots, they can be used to examine a single variable's distribution alone

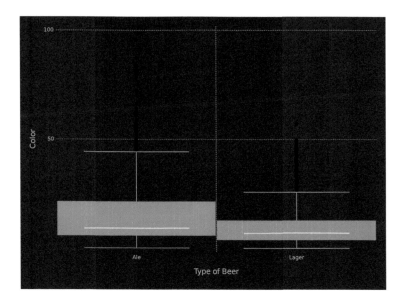

Figure 4.7 Boxplots of beer colour by type for the beer data, using two types.

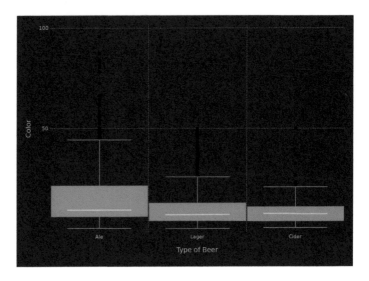

Figure 4.8 Boxplots of beer colour by type for the beer data, using three types.

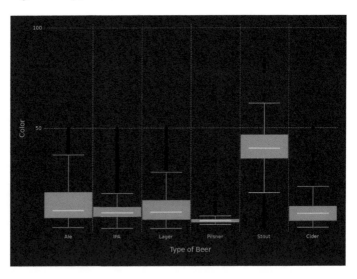

Figure 4.9 Boxplots of beer colour by type for the beer data, using six types.

or across categories. Because violin plots show the density of the data and not only summaries, they are particularly effective at discovering multimodal distributions, identifying clusters or bumps in the data, and identifying known distributions based on the shape of the density. The code to generate violin plots is detailed in the following code block. The only difference between this and the boxplot code is the `Geom` being used, highlighting the power and flexibility of the GoG framework.

```
## Violin plots
p_vio = plot(df_beer1,  x=:c3,  y =:pri_temp, Geom.violin,
    Guide.xlabel("Type of Beer"), Guide.ylabel("Primary Temperature"))

p_vio2 = plot(df_beer1,  x=:c3,  y =:color, Geom.violin,
    Guide.xlabel("Type of Beer"),  Guide.ylabel("Color"))

p_vio3 = plot(df_beer,  x=:c3,  y =:color, Geom.violin,
    Guide.xlabel("Type of Beer"),  Guide.ylabel("Color"))

p_vio4 = plot(df_beer,  x=:c6,  y =:color, Geom.violin,
    Guide.xlabel("Type of Beer"),  Guide.ylabel("Color"))
```

The resulting violin plots are displayed in Figures 4.10–4.13. Figure 4.10 supports the conclusions that can be drawn from the boxplot in Figure 4.7 and adds additional insight. Specifically, the majority of lagers have a lower colour rating which gradually tappers off in a series of humps as the density reaches 20. Colour ratings for ales, on the other hand, taper off smoothly to 20 units and have a larger portion of their observations in the right tail of the distribution, with a pronounced hump at 50 units. Figure 4.11 displays the violin plots for the primary brewing temperature of each beer. Lagers exhibit a clear bimodal pattern, indicating there are two groups of lagers being brewed at slightly different temperatures. Ales seem to be brewed mostly at temperatures consistent with the second group of lagers.

Another way to examine the distributions in data is to use quantile-quantile plots, also known as QQ-plots. They are used to compare the quantiles of one continuous variable against the quantiles of a known distribution or the quantiles of another variable. They are often used in statistical modelling to verify the distributional assumptions of different models. QQ-plots are scatterplots, where the points are evenly spaced quantiles of the data and/or distribution being examined. If a known distribution is used, the theoretical quantiles are plotted on one axis.

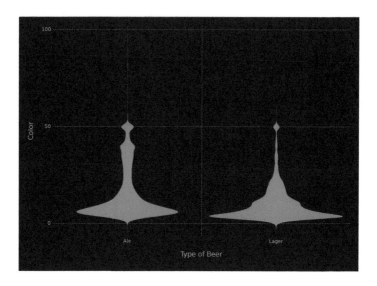

Figure 4.10 Violin plot of colour versus type for the beer data, using two types.

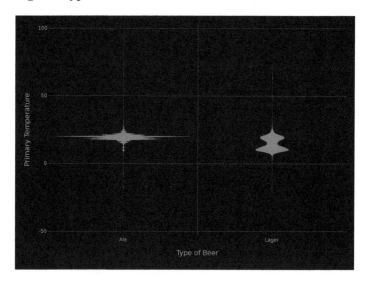

Figure 4.11 Violin plots of primary brewing temperature versus type for the beer data, using two types.

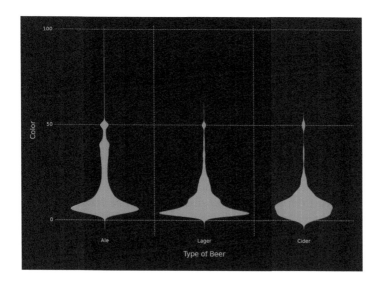

Figure 4.12 Violin plots of colour versus type for the beer data, using three types.

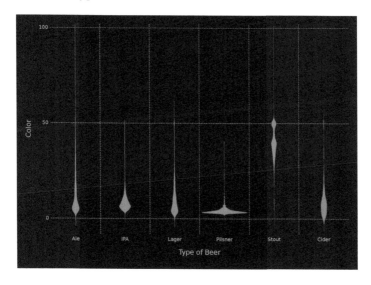

Figure 4.13 Violin plots of colour versus type for the beer data, using six types.

When the distributions of x and y align, the QQ-plot will have a linear trend along the $y = x$ line. If one distribution is a linear transformation of the other, the line will be straight but shifted away from the $y = x$ line. When the distributions differ, there will be a systematic pattern in the plot. If the points at the tails make the plot form an "S" shape, one of the distributions has heavier tails than the other. If the pattern is a "U" shape, one distribution is skewed in the direction of the curve of the "U".

The Julia code to generate the QQ-plots is given in the following code block. The first function call to `plot()` compares the simulated data in the `x3` column to a random sample of the same size drawn from a Pareto(2,1) distribution. The theoretical quantiles are drawn on the x-axis, as is convention for this style of QQ-plot. The `Stat.qq` statistic is used to generate the evenly spaced quantiles and they are rendered with the `point Geom`. The `abline Geom` is used to draw a line with a zero intercept and a slope of one.

```
p_qq1 = plot(df_1,  y=:x3,  x = rand(Pareto(2, 1), N), Stat.qq, Geom.point,
   Geom.abline(color="green", style=:dash), Guide.ylabel("Quantiles of X3"),
   Guide.xlabel("Quantiles of Pareto(2,1) Distribution"))

p_qq2 = plot(df_1, x=:x3, y = :x1, Stat.qq, Geom.point,
   Guide.xlabel("Quantiles of X3"),   Guide.ylabel("Quantiles of X1"))
```

Figure 4.14 displays a nice linear trend with some deviation at the right tail of the data. In practice, a QQ-plot will rarely be as well behaved as this because the `x3` variable was simulated from the same distribution as the theoretical quantiles. Figure 4.15 has a pronounced "U" shape, where the bottom of the "U" is in the upper left-hand side of the graph. This indicates that `x1` is skewed to the left of `x3`. This is expected given that `x1` is simulated from a Normal(2,1) distribution and its density is shifted to the left of the Pareto(2,1) distribution. As can be seen from these examples, important distributional information can be gleaned from a QQ-plot, which can help inform a data scientist's choice of modelling strategy.

An alternative way of visualizing distributions of data is the empirical cumulative distribution function (ECDF) plot. The ECDF is a function that estimates the fraction of observations below a given value of the measured data. Looking at Figure 4.16, we see that 50% of the data values are less than 10, 75% are less than 20, and 95% of values are below 42. The Julia code to generate the

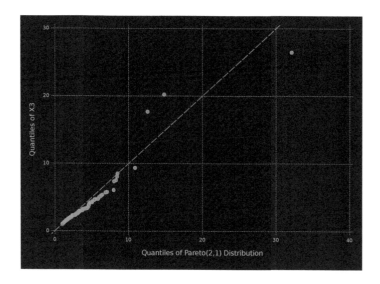

Figure 4.14 QQ-plot of simulated data and theoretical quantiles from a Pareto(2,1) distribution.

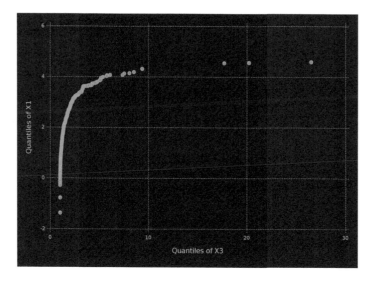

Figure 4.15 QQ-plot of simulated data from a Normal(2,1) and a Pareto(2,1) distributions.

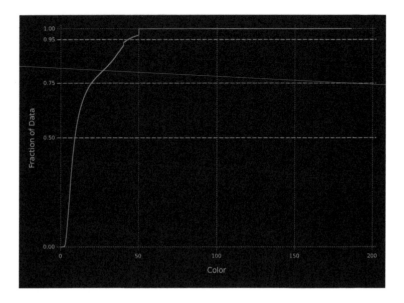

Figure 4.16 ECDF plot of the colour variable from the `beer` dataset.

ECDF plots is given in the following code block. We use the `ecdf()` function from the `StatsBase.jl` package to generate the ECDF plot for `colour` (Figure 4.17). `ecdf()` returns a function that must then be applied to the sample data to generate the ECDF values. The `step Geom` is used to draw the step function of the ECDF.

```
## single variable ecdf plot
ecdf_p1 = plot(df_ecdf, x = :color,
  y = ecdf(df_ecdf[:color])(df_ecdf[:color]),
  Geom.step,  yintercept=[0.5, 0.75, 0.95],
  Geom.hline(style = :dot, color = "gray"),
  Guide.yticks(ticks = [0,0.5, 0.75, 0.95, 1]),
  Guide.xlabel("Color"), Guide.ylabel("Fraction of Data"))

## ecdf plot by beer type
## dataframe of data and percentiles by labels
df_p = by(df_ecdf, :label, df -> DataFrame(
                 e_cdf = ecdf(df[:color])(df[:color]), col = df[:color]))

ecdf_p2 = plot(df_p, y = :e_cdf, x = :col, color = :label, Geom.step,
  yintercept=[0.5, 0.75, 0.95], Geom.hline(style = :dot, color = "gray"),
  Guide.yticks(ticks = [0,0.5, 0.75, 0.95, 1]), Guide.xlabel("Color"),
  Guide.ylabel("Fraction of Data"), Guide.colorkey(title = "Type of Beer"),
  Scale.color_discrete_manual("cyan", "darkorange", "magenta"))
```

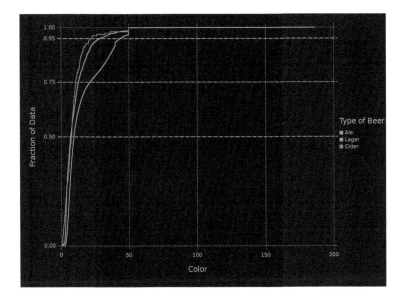

Figure 4.17 ECDF plot of the colour variable by beer type from the beer dataset.

ECDF plots really shine when multiple categories are compared across the same measured variable. They allow the viewer to quickly pick out where the distributions of the measured variable differ between categories. The previous code block details how to generate such a figure by plotting a separate ECDF plot for each type of beer. We use the split-apply-combine framework discussed in Section 3.5 to create the ECDF values for each beer label and then plot the data in the newly created dataframe. Figure 4.17 illustrates the results. We see that the ales are shifted to the right of the lagers and ciders, having larger values at every fraction of the data. This is supported by the box and violin plots discussed previously in the chapter.

4.4   VISUALIZING BIVARIATE DATA

We are often interested in the relationship between two (or more) quantities. When there are two quantities, bivariate plots can help visualize these relationships. Scatterplots are one of the most well-recognized bivariate plots, allowing the examination of two continuous measurements. They consist of two axes, one vertical and one horizontal, and show how much one measure affects the other. The relationships elucidated in a scatterplot can take any form, not just linear relationships. The QQ-plot discussed previously is an example of a scatterplot. Sometimes, a third variable can be represented on a scatterplot by adding colour coding or sizing the plotted symbols according to their value. The following code block shows how to make a simple scatterplot, which is then displayed in Figure 4.18. The x and y aesthetics point to two continuous measurements in the referenced dataframe df_1. The only other element need is the point Geom.

```
## basic scatterplot
sp_1 = plot(df_1, x=:x4, y=:y, Geom.point )

## scatterplot with color coding and a non-linear smoother
sp_2 = plot(df_1, x=:x4, y=:y, color = :x2, Geom.point,
    Scale.color_discrete_manual("red","purple","blue"),
    Geom.smooth(method=:loess,smoothing=0.5) )
```

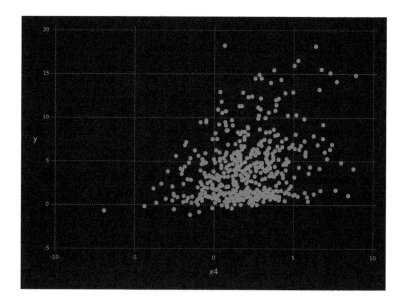

Figure 4.18 Scatterplot of y versus x4 from the simulated data.

Figure 4.18 is not very helpful in finding trends or patterns in the data. The second scatterplot is displayed in Figure 4.19, which adds a colour aesthetic for the categories in the x2 col-

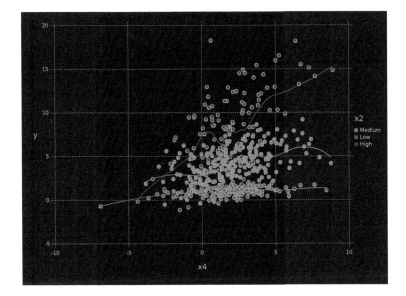

Figure 4.19 Scatterplot of y versus x4 with colour coding and a Loess smoother from the simulated data.

umn, a `colour_discrete_manual` Scale to add custom colours for the three levels of x2. Using the `smooth Geom`, a non-linear Loess smoother (Ruppert et al., 2003) is added to the plot to detect any non-linear trends within the three x2 groups. The degree of smoothing done by the Loess model can be adjusted by the smoothing argument, where smaller values result in less smoothing. The `smooth Geom` method argument can be changed to `:lm` to add linear trends to a plot.

The additional colour and non-linear trend lines visible in Figure 4.19 provide some interesting insights. There are three clear groups in the data, where each level has a more pronounced non-linear trend moving from Low to High. Without these two aspects, i.e., the colour and non-linear trend lines, we would not have gleaned these insights (cf. Figure 4.18).

A modern alternative to scatterplots are hexbin plots (Carr et al., 1987). They are particularly useful in situations where the dataset is very large. In these situations, scatterplots can become overwhelmed by the sheer number of points being displayed. Many of the points will overlap, to the extent that traditional remedies such as jittering and semi-transparency are ineffective. The overlap can misrepresent data, making it hard to visualize the trends therein. In situations such as this, hexbin plots can be invaluable. They group the two continuous measures into hexagonal bins which are coloured by the count or density of points in the bin and can be viewed as a form of bivariate histogram. The hexagonal bins are preferred over squares because they can be packed more densely into the same space and they do not draw the viewer's eyes to the vertical and horizontal grid line of the resulting plot, making for a more honest display of the data.

The following code block illustrates how to create a hexbin plot using Gadfly.jl. As is the case in most visualizations produced in a GoG framework, it requires only slight modifications to basic scatterplot code; here, the hexbin Geom is used in place of the points Geom. The colour scale for the counts is adjusted to vary from blue on the low end to orange at the high end.

```
## Hexbin plot

p_hexb = plot(df_beer, x=:color, y=:pri_temp, Geom.hexbin,
    Guide.xlabel("Color"), Guide.ylabel("Primary Temperature"),
    Scale.color_continuous(colormap=Scale.lab_gradient("blue", "white",
        "orange")))
```

We use the beer data to illustrate the hexbin plot because it has 75,000 observations. From Figure 4.20, there is a pronounced cluster of beers with low colour and moderate brewing temperatures. The pattern of increased counts extends horizontally across the plot, culminating in a cluster of beers having moderate brewing temperatures and high colour values. The horizontal pattern is also present at high and low temperatures.

Often data scientists would like to look at two categorical variables and how they relate to a given metric, be that a count of their levels or another quantitative measure associated with them. This can be accomplished using heat maps. Heat maps draw rectangular bins, where each bin corresponds to a combination of levels of the two categorical variables. The rectangles are coloured by the magnitude of the metric of interest. The following code block illustrates how to construct a heatmap for a correlation matrix, calculated from the continuous variables in the simulated data. The categorical variables are the variable names and the metric is the

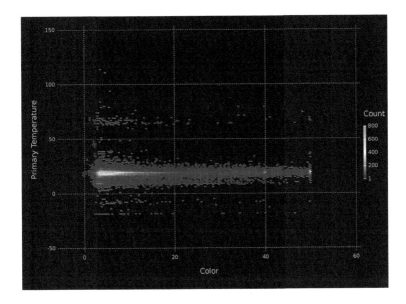

Figure 4.20   Hexbin plot of primary brewing temperature versus colour for the beer data.

correlation between them. The correlation matrix is extracted as a triangular matrix because it is symmetric around the diagonal.

```
## lower triangular correlation matrix
using LinearAlgebra

cor_mat = LowerTriangular(cor(
  convert(Array{Float64}, df_1[[:x1, :x3, :x4, :y]])
))

## convert cor_mat into a dataframe for GadFly
df_cor = DataFrame(cor_mat)
rename!(df_cor, [:x4 => :y, :x3 => :x4, :x2 => :x3])
df_cor = stack(df_cor)
df_cor[:variable2] = repeat(["x1", "x3", "x4", "y"], outer=4)

## plot the heatmap
p_hm = plot(df_cor, x=:variable2, y = :variable, color=:value,
  Geom.rectbin, Guide.colorkey(title = "Correlation"),
  Scale.color_continuous(colormap=Scale.lab_gradient("blue", "white",
  "orange"), minvalue=-1, maxvalue=1) )
```

The call to **plot()** assigns the correlation between the variables to the colour aesthetic and uses the **rectbin** Geom to draw the heatmap rectangles. The **colourkey** Geom is used to rename the colour map label and the `colour_continuous Scale` defines a custom colour gradient, moving from blue indicating a correlation of −1 through white denoting a correlation of 0 to orange designating a correlation of 1. Looking at the plot (Figure 4.21), the diagonal elements are orange as we would expect. The variable y is most strongly correlated with x1 and x4. This is as we would expect given both y and x2 are simulated from x1. The other variables have zero correlation as they are not related in any way.

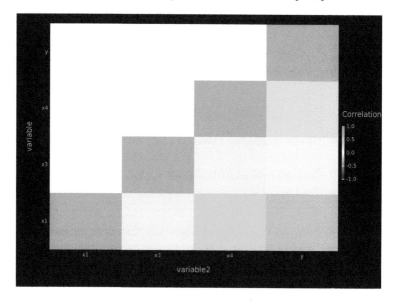

Figure 4.21 Heatmap of a correlation matrix from the simulated data.

Frequently data scientists want to graphically display a three-dimensional surface. One way to do this is to use contours to project it into the two-dimensional plane. `Gadfly.jl` can accomplish this by producing a contour plot using the `contour` geometry. A contour plot graphs two predictor variables, x and y, along their respective axes and draws the third dimension, z as contour lines in the xy-plane. Often, z is expressed as a function of the x and y values.

We detail how to generate a contour plot in the following code block. The code uses the covariance matrix S and the mean vector mu to plot the density function of the bivariate normal distribution. We use the `pdf()` and `MvNormal()` functions from the

`Distributions.jl` package to generate the graph's z values over a 100×100 grid of values from -10 to 10. The z argument in the `plot()` function call must produce one z value for each combination of the x and y values.

```
## Multivariate Normal parameters
N=1000
S = [2.96626 1.91; 1.91  2.15085]
mu = [0,5]

## contour plot of MVN over grid -10 to 10
ct_p = plot(z=(x,y) -> pdf(MvNormal(mu, S), [x, y]),
  x=range(-10, stop=10, length=N),
  y=range(-10, stop=10, length=N),
  Geom.contour)
```

The resulting plot is displayed in Figure 4.22. The large covariance between the x and y values results in the elliptical shape. The density values are displayed as a color gradient in the legend on the right.

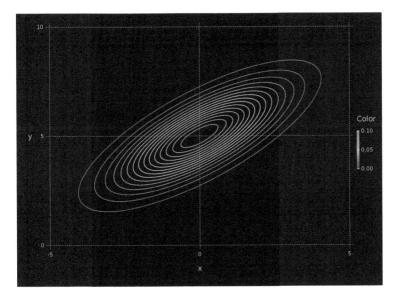

Figure 4.22 Contour plot for a bivariate normal distribution.

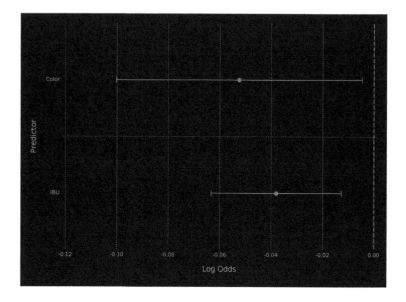

Figure 4.23 Error bars to summarize a logistic regression model run on the `beer` data.

4.5   ERROR BARS

In many data science problems, it is desirable to summarize model parameters and the uncertainty associated with these parameters. Often graphical displays are a better option for doing this than tables (Gelman et al., 2002). Parameter estimates and their uncertainty can be visualized using error bars. In `Gadfly.jl`, error bars are rendered using the `errorbar` `Geom`. It draws horizontal (x-axis) or vertical (y-axis) error bars and has arguments for both the minimum and maximum values of the bars. Error bars can also be coloured based on the values of a third variable.

Figure 4.23 summarizes the results of a logistic regression model with two predictor variables run on a random sample of the `beer` data. The parameter estimates and their 95% confidence intervals are plotted using the `point` and `errorbar` geometries. A vertical line is rendered at a log-odds of 0, which corresponds to the null hypothesis being tested for each parameter. The 95% confidence intervals do not overlap this line, indicating that the parameters are statistically significantly different from 0 at the 5% level.

4.6   FACETS

Facets are a generalization of scatterplot matrices, where a matrix of panels can be defined for different plotting types. They can be defined by column, row or both. Facets are particularly useful when it is desirable to stratify visualizations by one or two categorical variables. The subsets of data in each panel allow one to quickly compare patterns in different sections of the data. Creating facet columns helps make comparisons along the y-axis of subplots. This is useful in the supervised learning context when one wishes to visualize how the relationship between the outcome and an important predictor varies between facets. Row facets help make comparisons along the x-axis of plots and are especially useful when visualizing distributions. The following code block details how to produce a plot with row facets. The salient parts of the code are the `ygroup` aesthetic, which is bound to the dataframe column we wish to make the facet from, and the `subplot_grid` `Geom`. This `Geom` creates a matrix of panels and uses the `Geoms` in its argument along with the other `plot()` parameters to draw a unique plot in each panel.

```
p_hexb2 = plot(df_beer1, x=:color, y=:pri_temp, ygroup =:c3,
    Geom.subplot_grid(Geom.hexbin),
    Guide.xlabel("Color"), Guide.ylabel("Primary Temperature"),
    Scale.color_continuous(
        colormap=Scale.lab_gradient("blue", "white", "orange")
    )
)
```

The resulting plot is shown in Figure 4.24. By rendering the plots in rows, many of the features we identified in Figures 4.10 and 4.11 are clearly visible. The bimodal pattern in the distribution of the primary brewing temperature of lager can be identified along with the long and wide tail shape in the distribution of ale colours.

4.7   SAVING PLOTS

Saving plots is trivial using Gadfly.jl. The `draw()` function is used to save the plots. It is beneficial to assign a variable name to the plots to be saved to reference it in the `draw()` function. In the previous code block, the variable `p_hexb2` can be used to reference the Julia object that renders Figure 4.24 in this way. The `draw()` function writes SVG output by default. Additional backends can be accessed through the `Cairo.jl` package and include PNG, PDF

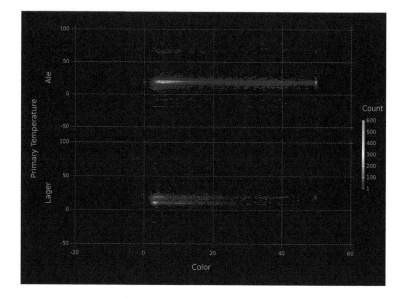

Figure 4.24 Facet plot for the beer data.

and PS. The following code block shows how to write the p_1 plot object to the dswj_plot1.eps postscript file.

```
draw(PS("dswj_plot1.eps", 197mm, 141mm), p_1)
```

## CHAPTER 5

Supervised Learning

S UPERVISED LEARNING uses labelled data to make predictions about unlabelled data. One defining characteristic of supervised learning is that the goal is prediction, rather than modelling or inference. A basic nomenclature, notation and framework for supervised learning is laid down before cross-validation is introduced. After that, a simple and yet quite effective supervised learning approach, i.e., K-nearest neighbours classification, is discussed. The famous CART (classification and regression trees) framework is discussed next, followed by the bootstrap. The bootstrap is preparatory for the use of CART for ensemble learning; specifically for random forests. The chapter concludes with another ensemble learning approach, gradient boosting using XGBoost.

5.1   INTRODUCTION

Consider the situation where there is an outcome variable Y that we want to predict based on several predictor variables $X_1, \ldots, X_p$, which we express as a vector $\mathbf{X} = (X_1, \ldots, X_p)'$. Suppose that we observe n pairs $(y_1, \mathbf{x}_1), \ldots, (y_n, \mathbf{x}_n)$ and the objective is to learn the values of "new" $y_{n+1}, \ldots, y_m$ for corresponding values of the predictor variables, i.e., for $\mathbf{x}_{n+1}, \ldots, \mathbf{x}_m$. Broadly speaking, one can use the observed pairs $(y_1, \mathbf{x}_1), \ldots, (y_n, \mathbf{x}_n)$ to build a model of the form

$$y_i = f(\mathbf{x}_i) + e_i, \tag{5.1}$$

where $f(\mathbf{x})$ is a function, sometimes called the learner or predictor, that takes a value $\mathbf{x}_i$ and returns a prediction for y_i, and e_i is the associated error. Rearranging (5.1) gives

$$e_i = y_i - f(\mathbf{x}_i), \tag{5.2}$$

for $i = 1, \ldots, n$, which is the error incurred by using $f(\mathbf{x}_i)$ to predict y_i. Some function of (5.2) can then be used to assess how well the learner performs before we apply it to predict y_i for $i = n + 1, \ldots, m$. Within the supervised learning paradigm, $(y_1, \mathbf{x}_1), \ldots, (y_n, \mathbf{x}_n)$ can be thought of as the *labelled* data because each value $\mathbf{x}_i$ has a corresponding "label" y_i. Accordingly, the $\mathbf{x}_{n+1}, \ldots, \mathbf{x}_m$ are considered *unlabelled* data because there are no labels, i.e., no values $y_{n+1}, \ldots, y_m$, attached to the values $\mathbf{x}_{n+1}, \ldots, \mathbf{x}_m$.

The very general modelling approach described by (5.1) and (5.2) is quite common in data science and statistics. For example, in a regression setting with $y_i \in \mathbb{R}$ there are a few functions of (5.2) that are commonly used to assess how well the learner performs. These include the mean squared error (MSE)

$$\frac{1}{n} \sum_{i=1}^{n} e_i^2 = \frac{1}{n} \sum_{i=1}^{n} (y_i - f(\mathbf{x}_i))^2, \tag{5.3}$$

for $i = 1, \ldots, n$, the root mean squared error (RMSE)

$$\sqrt{\frac{1}{n} \sum_{i=1}^{n} e_i^2} = \sqrt{\frac{1}{n} \sum_{i=1}^{n} (y_i - f(\mathbf{x}_i))^2}, \tag{5.4}$$

for $i = 1, \ldots, n$, and the median absolute error (MAE)

$$\text{median}_{i=1,\ldots,n} |e_i| = \text{median}_{i=1,\ldots,n} |y_i - f(\mathbf{x}_i)|. \tag{5.5}$$

In a binary classification setting with $y_i \in \{0, 1\}$ and $f(\mathbf{x}_i) \in \{0, 1\}$, one could use the misclassification rate to assess how well the predictor $f(\mathbf{x}_i)$ performs. In this case, i.e., with $y_i \in \{0, 1\}$ and $f(\mathbf{x}_i) \in \{0, 1\}$, the misclassification rate can be written

$$\frac{1}{n} \sum_{i=1}^{n} |e_i| = \frac{1}{n} \sum_{i=1}^{n} |y_i - f(\mathbf{x}_i)|, \tag{5.6}$$

for $i = 1, \ldots, n$. Note that (5.6) would cease to be effective in a binary classification setting with $y_i \in \{-1, 1\}$ and $f(\mathbf{x}_i) \in \{-1, 1\}$, for instance, or in the multinomial case, i.e., where there are more than two classes. Equation (5.6) can be formulated more generally for the misclassification rate:

$$\mathcal{E}_{\text{labelled}} = \frac{1}{n} \sum_{i=1}^{n} \mathbb{I}(y_i \neq f(\mathbf{x}_i)), \tag{5.7}$$

where $\mathbb{I}(y_i \neq f(\mathbf{x}_i))$ is an indicator function, i.e.,

$$\mathbb{I}(y_i \neq f(\mathbf{x}_i)) = \begin{cases} 1 & \text{if } y_i \neq f(\mathbf{x}_i), \\ 0 & \text{if } y_i = f(\mathbf{x}_i). \end{cases}$$

For some, functions such as the misclassification rate are considered too coarse for model selection. A common alternative for binary classification is the binomial deviance, or simply deviance, which can be written

$$\mathcal{E}_{\text{labelled}} = -2 \sum_{i=1}^{n} y_i \log(\hat{\pi}_i) + (1 - y_i) \log(1 - \hat{\pi}_i), \tag{5.8}$$

where $\hat{\pi}_i$ is the predicted probability of y_i being equal to 1 and is generated from $f(\mathbf{x}_i)$. The quantity in (5.8) is negative two times the maximized binomial log-likelihood (McCullagh and Nelder, 1989). The deviance is a proper scoring rule (Gneiting and Raftery, 2007), which helps ensure that the predicted probabilities fit the data well and the learner is well calibrated to the data. Many other proper scoring rules exist, such as the Brier score (Brier, 1950). Binary classification models used herein are trained on the deviance and we report it along with the misclassification rate.

Hereafter, we shall use the notation $\mathcal{E}_{\text{labelled}}$ to denote a generic error function for the labelled data. In general, $\mathcal{E}_{\text{labelled}}$ will take the form

$$\mathcal{E}_{\text{labelled}} = g((y_1, f(\mathbf{x}_1)), \dots, (y_n, f(\mathbf{x}_n))), \tag{5.9}$$

where $g((y_1, f(\mathbf{x}_1)), \dots, (y_n, f(\mathbf{x}_n)))$ is a function that reflects how close $f(\mathbf{X})$ is to Y for the training data; such a function is sometimes called a loss function or an error function.

When doing supervised learning, the choice of error function is very important. In regression problems, the RMSE (5.4) is the standard choice; however, the RMSE is highly sensitive to outliers because it squares the errors e_i. Instead of using RMSE, we train our regression models herein to minimize the MAE (5.5). The MAE is insensitive to outliers in the outcome because the median is not affected by values in the tails of the error distribution.

The learning approach described thus far has some important limitations:

1. We have said nothing about how the predictor $f(\mathbf{x})$ is constructed.

2. Constructing $f(\mathbf{x})$ based on $(y_1, \mathbf{x}_1), \ldots, (y_n, \mathbf{x}_n)$ and also assessing the error $\mathcal{E}_{\text{labelled}}$ based on $(y_1, \mathbf{x}_1), \ldots, (y_n, \mathbf{x}_n)$ will tend to underestimate $\mathcal{E}_{\text{unlabelled}}$, i.e., the error that one would see when $f(\mathbf{x})$ is applied to the unlabelled data. In other words, $\mathcal{E}_{\text{labelled}}$ is a biased estimator of $\mathcal{E}_{\text{unlabelled}}$.

With respect to the above list item 1., it is not possible to detail exactly how the learner $f(\mathbf{x})$ is constructed because this will depend on the learning method used. However, we can say a little more about how the learner $f(\mathbf{x})$ is constructed in general and doing so also serves as a response to list item 2. Rather than using all of the labelled data to build the learner $f(\mathbf{x})$, the labelled data are partitioned into a *training set* and a *test set*. Then, the learner $f(\mathbf{x})$ is constructed based on the training set and the error is assessed based on the test set. Of course, it is reasonable to expect the test error $\mathcal{E}_{\text{test}}$ to be more reflective of $\mathcal{E}_{\text{unlabelled}}$ than either $\mathcal{E}_{\text{labelled}}$ or the training error $\mathcal{E}_{\text{training}}$ would be.

While we have gone some way towards addressing list item 1., and seemingly addressed list item 2., we have created a third problem:

3. If we overfit $f(\mathbf{x})$ to the training data, then it may not perform well on the test data and/or the unlabelled data even though the training error $\mathcal{E}_{\text{training}}$ may be very small.

The existence of list item 3. as a problem is, in the first place, reflective of the fact that the goal of supervised learning is prediction. This is a key point of difference with traditional statistical procedures, where the goal is modelling or inference — while inference can be taken to include prediction, prediction is not the goal of inference. To solve the problem described in list item 3., we need to move away from thinking only about an error and consider an error together with some way to prevent or mitigate overfitting. Finally, it is worth noting that some learning approaches are more prone to overfitting than others; see Section 5.8 for further discussion.

5.2 CROSS-VALIDATION

5.2.1 Overview

We have seen that the training set is used to construct a learner $f(\mathbf{x})$. Now, consider how this is done. One approach is to further partition the training set into two parts, and to use one part to build lots of learners and the other to choose the "best" one. When this learning paradigm is used, the part that is used to build lots of learners is called the training set, and the part used to choose the best learner is called the validation set. Accordingly, such a learning framework is called a training-validation-test framework. Similarly, the paradigm discussed in Section 5.1 can be thought of as a training-test framework.

One longstanding objection to the training-validation-test framework is that a large dataset would be required to facilitate separate training and validation sets. There are several other possible objections, one of which centres around the sensitivity of the results to the training-validation split, which will often be equal or at least close, e.g., a 40-40-20 or 50-40-10 training-validation-test split might be used. Given the increasing ease of access to large datasets, the former objection is perhaps becoming somewhat less important. However, the latter objection remains quite valid. Rather than using the training-validation-test framework, the training-test paradigm can be used with cross-validation, which allows both training and validation within the training set.

5.2.2 K-Fold Cross-Validation

K-fold cross-validation partitions the training set into K (roughly) equal parts. This partitioning is often done so as to ensure that the y_i are (roughly) representative of the training set within each partition — this is known as stratification, and it can also be used during the training-test split. In the context of cross-validation, stratification helps to ensure that each of the K partitions is in some sense representative of the training set. On each one of K iterations, cross-validation proceeds by using $K - 1$ of the folds to construct the learner and the remaining fold to compute the error. Then, after all iterations are completed, the K errors are combined to give the cross-validation error.

The choice of K in K-fold cross-validation can be regarded as finding a balance between variance and bias. The variance-bias tradeoff is well known in statistics and arises from the fact that, for an estimator of a parameter,

$$\text{MSE} = \text{Variance} + \text{Bias}^2.$$

Returning to K-fold cross-validation, choosing $K = n$, sometimes called leave-one-out cross-validation, results is excellent (i.e., low) bias but high variance. Lower values of K lead to more bias but less variance, e.g., $K = 10$ and $K = 5$ are popular choices. In many practical examples, $K = n$, $K = 10$ and $K = 5$ will lead to similar results.

The following code block illustrates how one could write their own cross-validation routine in Julia. Note that this serves, in part, as an illustration of how to write code in Julia — there are, in fact, cross-validation functions available within Julia (see, e.g., Section 5.7). The first function, cvind() breaks the total number of rows in the training set, i.e., N, into k nearly equal groups. The folds object returned by the function is an array of arrays. Each sub-array contains gs randomly shuffled row indices. The last sub-array will be slightly larger than gs if k does not divide evenly into N. The second function, kfolds(), divides the input data dat into k training-test dataframes based on the indices generated by cvind(). It returns a dictionary of dictionaries, with the top level key corresponding to the cross-validation fold number and the subdictionary containing the training and test dataframes created from that fold's indices. This code expects that dat is a dataframe and would not be efficient for dataframes with large N. In this scenario, the indices should be used to subset the data and train the supervised learners in one step, without storing all the training-test splits in a data structure.

```
using Random
Random.seed!(35)

## Partition the training data into K (roughly) equal parts
function cvind(N, k)
        gs = Int(floor(N/k))
        ## randomly shuffles the indices
    index = shuffle(collect(1:N))
    folds = collect(Iterators.partition(index, gs))
        ## combines and deletes array of indices outside of k
    if length(folds) > k
        folds[k] = vcat(folds[k], folds[k+1])
        deleteat!(folds, k+1)
    end
    return folds
end

## Subset data into k training/test splits based on indices
## from cvind
function kfolds(dat, ind, k)
    ## row indices for dat
    ind1 = collect(1:size(dat)[1])
    ## results to return
    res = Dict{Int, Dict}()
    for i = 1:k
        ## results for each loop iteration
        res2 = Dict{String, DataFrame}()
        ## indices not in test set
        tr = setdiff(ind1, ind[i])
```

```
        ## add results to dictionaries
        push!(res2, "tr"=>dat[tr,:])
        push!(res2, "tst"=>dat[ind[i],:])
        push!(res, i=>res2)
    end
    return res
end
```

5.3 K-NEAREST NEIGHBOURS CLASSIFICATION

The k-nearest neighbours (kNN) classification technique is a very
simple supervised classification approach, where an unlabelled observation is classified based on the labels of the k closest labelled
points. In other words, an unlabelled observation is assigned to the
class that has the most labelled observations in its neighbourhood
(which is of size k). The value of k needs to be selected, which can
be done using cross-validation, and the value of k that gives the
best classification rate under cross-validation is generally chosen.
Note that, if different k give similar or identical classification rates,
the smaller value of k is usually chosen.

The next code block shows how the kNN algorithm could be
implemented in Julia. The knn() function takes arrays of numbers
for the existing data and labels, the new data and a scalar for k. It
uses Manhattan distance to compare each row of the existing data
to a row of the new data and returns the majority vote of the labels
associated with the smallest k distances between the samples. The
maj_vote() function is used to calculate the majority vote, and
returns the class and its proportion as an array.

```
## majority vote
function maj_vote(yn)
    ## majority vote
    cm = countmap(yn)
    mv = -999
    lab = nothing
    tot = 1e-8
    for (k,v) in cm
        tot += v
        if v > mv
            mv = v
            lab = k
        end
    end
    prop = /(mv, tot)
    return [lab, prop]
end

## KNN label prediction
```

```
function knn(x, y, x_new, k)
    n,p = size(x)
    n2,p2 = size(x_new)
    ynew = zeros(n2,2)

    for i in 1:n2 ## over new x_new
        res = zeros(n,2)
        for j in 1:n ## over x
            ## manhattan distance between rows - each row is a subject
            res[j,:] = [j , cityblock(x[j,:], x_new[i,:])] #cityblock
        end
        ## sort rows by distance and index - smallest distances
        res2 = sortslices(res, dims = 1, by = x -> (x[2], x[1]))
        ## get indices for the largest k distances
        ind = convert(Array{Int}, res2[1:k, 1])
        ## take majority vote for the associated indices
        ynew[i,:] = maj_vote(y[ind])
    end
    ## return the predicted labels
    return ynew
end

## Returns the missclassification rate
function knnmcr(yp, ya)
    disagree = Array{Int8}(ya .!= yp)
    mcr = mean(disagree)
    return mcr
end
```

An illustration of kNN is given in Figure 5.1, where $k = 3$ and the neighbourhoods for each class are clearly visible. An unlabelled observation in one of the red neighbourhoods will be labelled as belonging to the red class, and similarly for the blue neighbourhood.

The next code block details how we can use the cross-validation functions already described to do 5-fold cross-validation for a kNN learner. Simulated data, df_knn which we mean centre and scale to have standard deviation one (not shown), is divided into 5 training-test splits using the cvind() and kfold() functions. We loop through the 15 values of k and, for each one, calculate the mean misclassification rate and its standard error over the 5 folds.

```
## Simulated data
df_3 = DataFrame(y = [0,1], size = [250,250], x1 =[2.,0.], x2 =[-1.,-2.])

df_knn =by(df_3, :y) do df
    DataFrame(x_1 = rand(Normal(df[1,:x1],1), df[1,:size]),
    x_2 = rand(Normal(df[1,:x2],1), df[1,:size]))
end

## set up parameters for cross-validation
N = size(df_knn)[1]
kcv = 5
```

```
## generate indices
a_ind = cvind(N, kcv)

## generate dictionary of dataframes
d_cv = kfolds(df_knn, a_ind, kcv)

## knn parameters
k = 15
knnres = zeros(k,3)

## loop through k train/test sets and calculate error metric
for i = 1:k
  cv_res = zeros(kcv)
  for j = 1:kcv
      tr_a = convert(Matrix, d_cv[j]["tr"][[:x_1, :x_2]])
      ytr_a = convert(Vector, d_cv[j]["tr"][:y])
      tst_a = convert(Matrix, d_cv[j]["tst"][[:x_1, :x_2]])
      ytst_a = convert(Vector, d_cv[j]["tst"][:y])
      pred = knn(tr_a, ytr_a, tst_a, i)[:,1]
      cv_res[j] = knnmcr(pred, ytst_a)
  end
  knnres[i, 1] = i
  knnres[i, 2] = mean(cv_res)
  knnres[i, 3] = /(std(cv_res), sqrt(kcv))
end
```

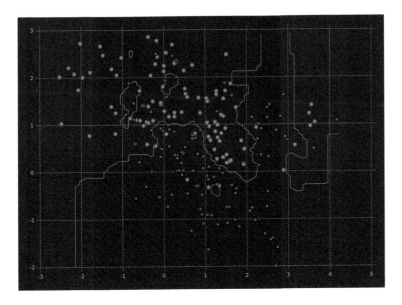

Figure 5.1 An illustration of kNN, for $k = 3$, using simulated data.

The cross-validation results are depicted in Figure 5.2. This plot shows an initial improvement in misclassification rate up to $k = 5$, with the smallest CV misclassification rate occurring at $k = 15$. For values of k from five to 15, the values oscillate up and down but remain within one standard error of $k = 15$. Based on these results, we would select $k = 5$, as it is the simplest learner that attains close to the minimum misclassification rate.

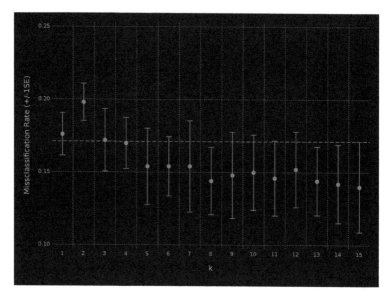

Figure 5.2 Results of 5-fold CV for 15 values of k using kNN on simulated data, where the broken line indicates one standard deviation above the minimum CV misclassification rate.

5.4 CLASSIFICATION AND REGRESSION TREES

5.4.1 Overview

Classification and regression trees (CART) were introduced by Breiman et al. (1984). They are relatively easy to understand and explain to a non-expert. The fact that they naturally lend themselves to straightforward visualization is a major advantage.

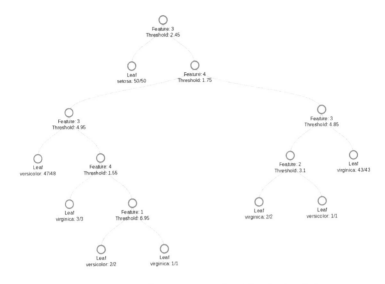

Figure 5.3 Classification tree for the iris data.

5.4.2 Classification Trees

First, consider a classification tree, which starts at the top (root) and recursively partitions data based on the best splitter. A classification tree for the famous iris data is shown in Figure 5.3, where the leaves report the result (i.e., one misclassification in total). There are eight leaves: one where irises are classified as *setosa*, three for *versicolor*, and four for *virginica*. The representation of the tree in Figure 5.3 gives the value of the feature determining the split and the fraction of observations classified as the iris type in each leaf. Features are numbered and correspond to sepal length, sepal width, petal length, and petal width, respectively.

As we saw with the iris tree (Figure 5.3), a splitter is a variable (together with a rule), e.g.,

$$\text{petal width} < 1.75.$$

What it means to be the "best" splitter will be discussed, *inter alia*, shortly. An in-depth discussion on pruning is probably not helpful at this stage; however, there are many sources that contain such details (e.g., Breiman et al., 1984; Hastie et al., 2009). A classification tree is also known as a decision tree. Consider the notation of Hastie et al. (2009), so that a node m represents a region

R_m with N_m observations. When one considers the relationship between a tree and the space of observations, this is natural. Before proceeding, consider a tree built using two variables from the iris data — two variables so that we can visualize the partitioning of the space of observations (Figure 5.4).

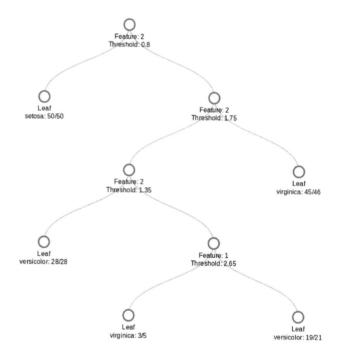

Figure 5.4 Classification tree for the Iris data using only two variables: petal length and petal width.

An alternative way to visualize Figure 5.4 is via a scatterplot with partitions (Figure 5.5). Each one of the five regions shown in Figure 5.5 corresponds to a leaf in the classification tree shown in Figure 5.4.

Recall that a node m represents a region R_m with N_m observations. Again, remaining with the notation of Hastie et al. (2009), the proportion of observations from class g in node m is

$$\hat{p}_{mg} = \frac{1}{N_m} \sum_{\mathbf{x}_i \in R_m} \mathbb{I}(y_i = g).$$

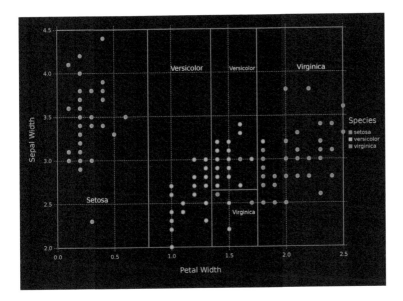

Figure 5.5 Partitioned scatterplot for the iris data.

All observations in node m are classified into the class with the largest proportion of observations; in other words, the majority class:

$$g^* = \arg\max_k \hat{p}_{mg}.$$

This is all just a formalization of what is quite clear by inspection of the tree.

Consider how to come up with splits (or splitters) in the first place. One approach is to base them on the misclassification error, i.e.,

$$1 - \hat{p}_{mg^*}.$$

Splits can also be based on the Gini index, i.e.,

$$\sum_{g=1}^{G} \hat{p}_{mg}(1 - \hat{p}_{mg})$$

or the cross-entropy (also called information or deviance), i.e.,

$$-\sum_{g=1}^{G} \hat{p}_{mg} \log \hat{p}_{mg}. \tag{5.10}$$

Note that when used in this way, misclassification error, Gini index, and cross-entropy are referred to as impurity measures. In this vernacular, we want nodes that are as pure as possible. The misclassification error is generally not the impurity measure of choice because it is not differentiable and so less amenable to numerical methods. Accordingly, the Gini index or cross-entropy is typically used.

5.4.3 Regression Trees

Now, consider regression trees. They proceed in an analogous fashion to classification trees. However, we now have a regression problem as opposed to a classification problem. The notion of impurity is somewhat more straightforward here and common choices for an impurity measure include the RMSE (5.4) and the MAE (5.5). Splitting requires a little thought. The problem can be formulated as choosing a value v and a variable X_j to split the (training) data into the regions

$$R_1(X_j, v) = \{\mathbf{X} \mid X_j < v\} \quad \text{and} \quad R_2(X_j, v) = \{\mathbf{X} \mid X_j \geq v\}.$$

The choice of j and v is made to minimize some loss function, such as

$$\sum_{i:\mathbf{x}_i \in R_1(j,v)} (y_i - \hat{y}_{R_1})^2 + \sum_{i:\mathbf{x}_i \in R_2(j,v)} (y_i - \hat{y}_{R_2})^2,$$

where $\hat{y}_{R_k}$ is the mean of the (training) observations in R_k, for $k = 1, 2$. Further splits continue in this fashion.

Consider the following code block. It uses the `DecisionTree.jl` package to create a regression tree from the `food` data. The leaves of the tree are merged if the purity of the merged leaf is at least 85%. The resulting regression tree has a depth of 8 and 69 leaves. A tree of this size can be hard to interpret. The `DecisionTree.jl` package can print out a textual representation of the tree but this leaves much to be desired. At the time of writing, `DecisionTree.jl` does not support graphical visualizations and plotting a decision tree in `Gadfly.jl` is not straightforward. We used an unofficial Julia package called `D3DecisionTrees.jl` to make interactive visualizations of our trees. They render in a web browser and allow the user to click through the nodes and branches to examine subsections of their tree. A static representation of one of these visualizations is used to partially visualize the tree built from the `food` data (Figure 5.6). Based on an 80-20

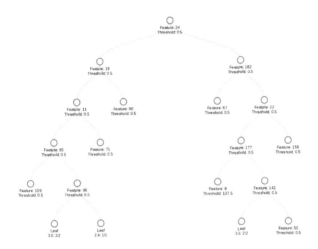

Figure 5.6 Regression tree built from the food data.

train test split of the data, the tree has a test set MAE of 0.82 for predicting student GPA.

```
using DecisionTree, Random
Random.seed!(35)

## food training data from chapter 3
y = df_food[:gpa]
tmp = convert(Array, df_food[setdiff(names(df_food), [:gpa])] )
xmat = convert(Array{Float64}, collect(Missings.replace(tmp, 0)))
names_food = setdiff(names(df_food), [:gpa])

# defaults to regression tree if y is a Float array
model = build_tree(y, xmat)

# prune tree: merge leaves having >= 85% combined purity (default: 100%)
modelp = prune_tree(model, 0.85)

# tree depth is 9
depth(modelp)

# print the tree to depth 3
print_tree(modelp, 3)
#=
Feature 183, Threshold 0.5
L-> Feature 15, Threshold 0.5
    L-> Feature 209, Threshold 0.5
        L->
        R->
    R-> Feature 209, Threshold 0.5
        L->
        R->
R-> Feature 15, Threshold 0.5
```

```
     L-> Feature 8, Threshold 0.5
        L->
        R->
    R-> Feature 23, Threshold 0.5
        L->
        R->
=#

# print the variables in the tree
for i in [183, 15, 209, 8, 23]
  println("Variable number: ", i, " name: ", names_food[i])
end

#=
Variable number: 183 name: self_perception_wgt3
Variable number: 15 name: cal_day3
Variable number: 209 name: waffle_cal1315
Variable number: 8 name: vitamins
Variable number: 23 name: comfoodr_code11
=#
```

5.4.4 Comments

Trees have lots of advantages. For one, they are very easy to understand and to explain to non-experts. They mimic decision processes, e.g., of a physician perhaps. They are very easy (natural, even) to visualize. However, they are not as good as some other supervised learning methods. Ensembles of trees, however, can lead to markedly improved performance. Bagging, random forests and gradient boosting are relatively well-known approaches for combining trees. However, we need to discuss ensembles and the bootstrap first.

5.5 BOOTSTRAP

The introduction of the bootstrap by Efron (1979) is one of the single most impactful works in the history of statistics and data science. Use $\mathcal{Q}$ to denote the sample $x_1, x_2, \ldots, x_n$. Note that, for the purposes of this section, taking x_i to be univariate will simplify the explanation. Now, $\mathcal{Q}$ can be considered a set, i.e.,

$$\mathcal{Q} = \{x_1, x_2, \ldots, x_n\},$$

and is sometime called an ensemble. Suppose we want to construct estimator $\hat{\theta}$ based upon that sample, and we are interested in the bias and standard error of $\hat{\theta}$. A resampling technique can be used, where we draw samples from an ensemble that is itself a sample, e.g., from Q. There are a variety of resampling techniques available; the most famous of these is called the bootstrap.

Before we look at an illustration of the bootstrap, we need to consider the plug-in principle. Let $X_1, X_2, \ldots, X_n$ be iid random variables. Note that the cumulative distribution function

$$G(x) = \frac{1}{n} \sum_{i=1}^{n} \mathbb{I}(X_i \leq x)$$

defines an empirical distribution. The plug-in estimate of the parameter of interest θ_F is given by

$$\hat{\theta} = \theta_{\hat{F}},$$

where $\hat{F}$ is an empirical distribution, e.g., summary statistics are plug-in estimates.

Efron (2002) gives an illustration of the bootstrap very similar to what follows here. Suppose that the data are a random sample from some unknown probability distribution F, θ is the parameter of interest, and we want to know $\mathrm{SE}_F(\hat{\theta})$. We can compute $\mathrm{SE}_{\hat{F}}(\hat{\theta})$, where $\hat{F}$ is the empirical distribution of F, as follows. Suppose we have an ensemble

$$Q = \{x_1, x_2, \ldots, x_n\}.$$

Sample with replacement from Q, n times, to get a bootstrap sample

$$Q^* = \{x_1^*, x_2^*, \ldots, \ldots, x_n^*\}$$

and then compute $\hat{\theta}^*$ based on Q^*. Repeat this process M times to obtain values

$$\hat{\theta}^*(1), \hat{\theta}^*(2), \ldots, \hat{\theta}^*(M)$$

based on bootstrap samples

$$Q^*(1), Q^*(2), \ldots, Q^*(M).$$

Then,

$$\widehat{\mathrm{SE}}_{\mathrm{boot}} = \sqrt{\frac{1}{M-1} \sum_{j=1}^{M} (\hat{\theta}^*(j) - \hat{\theta}^*(\cdot))^2}, \qquad (5.11)$$

where

$$\hat{\theta}^*(\cdot) = \frac{1}{M} \sum_{i=1}^{M} \hat{\theta}^*(i).$$

As $M \to \infty$,

$$\widehat{SE}_{boot} \to \widehat{SE}_{\hat{F}},$$

where $\widehat{SE}_{boot}$ and $\widehat{SE}_{\hat{F}}$ are non-parametric bootstrap estimates because they are based on $\hat{F}$ rather than F. Clearly, we want M to be as large as possible, but how large is large enough? There is no concrete answer but experience helps one get a feel for it.

```julia
using StatsBase, Random, Statistics
Random.seed!(46)

A1 = [10,27,31,40,46,50,52,104,146]
median(A1)
# 46.0

n = length(A1)
m = 100000
theta = zeros(m)

for i = 1:m
  theta[i] = median(sample(A1, n, replace=true))
end

mean(theta)
# 45.767
std(theta)
#12.918

## function to compute the bootstrap SE, i.e., an implementation of (5.11)
function boot_se(theta_h)
  m = length(theta_h)
  c1 = /(1, -(m,1))
  c2 = mean(theta_h)
  res = map(x -> (x-c2)^2, theta_h)
  return(sqrt(*(c1, sum(res))))
end

boot_se(theta)
## 12.918
```

Now, suppose we want to estimate the bias of $\hat{\theta}$ given an ensemble $\mathcal{Q}$. The bootstrap estimate of the bias, using the familiar notation, is

$$\widehat{Bias}_{boot} = \hat{\theta}^*(\cdot) - \hat{\theta},$$

where $\hat{\theta}$ is computed based the empirical distribution $\hat{F}$.

```julia
## Bootstrap bias
-(mean(theta), median(A1))
#-0.233
```

In addition to estimating the standard error and the bias, the bootstrap can be used for other purposes, e.g., to estimate confidence intervals. There are a number of ways to do this, and there is a good body of literature around this topic. The most straightforward method is to compute bootstrap percentile intervals.

```
## 95% bootstrap percentile interval for the median
quantile(theta, [0.025,0.975])

# 2-element Array{Float64,1}:
#   27.0
#  104.0
```

With the very brief introduction here, we have only scratched the surface of the bootstrap. There are many useful sources for further information on the bootstrap, including the excellent books by Efron and Tibshirani (1993) and Davison and Hinkley (1997).

5.6 RANDOM FORESTS

Consider a regression tree scenario. Now, generate M bootstrap ensembles from the training set and use $\hat{f}^m(\mathbf{x})$ to denote the regression tree learner trained on the mth bootstrap ensemble. Averaging these predictors gives

$$\hat{f}_{\text{bag}}(\mathbf{x}) = \frac{1}{M} \sum_{m=1}^{M} \hat{f}^m(\mathbf{x}).$$

This is called bootstrap aggregation, or bagging (Breiman, 1996). Although it seems very simple, bagging tends to result in a better learner than a single tree learner. Note that, with bagging, the M trees are not pruned. Bagging also gives a method for estimating the (test) error, known as the out-of-bag error. When we create a bootstrap ensemble, some observations will be included more than once and some will be left out altogether — the left-out observations are said to be out-of-bag. Typically, around 37% of observations will be out-of-bag for a given tree; to see this, consider that

$$\left(\frac{n-1}{n} \right)^n \tag{5.12}$$

is the probability that a single observation is omitted from a bootstrap sample (of size n) and then note how (5.12) behaves as n increases (Figure 5.7).

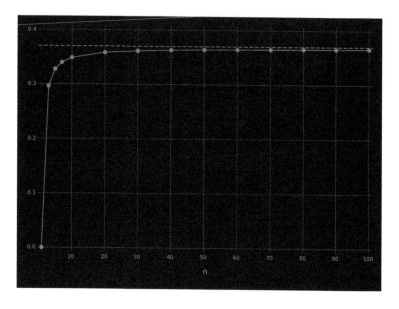

Figure 5.7 Illustration of the value of (5.12) as n increases, with a broken horizontal line at 0.37.

The response for the ith observation can be predicted using the trees for which it was out-of-bag. These can then be averaged to give a prediction for the ith observation. This is sometimes used as an estimate of the test error; however, we will not follow that approach herein.

While bagging can (perhaps greatly) improve regression tree performance, it comes at the expense of the easy visualization and easy interpretability enjoyed by a single regression tree. However, we can evaluate the importance of each predictor variable by considering the extent to which it decreases some cost function, e.g., the RMSE, averaged over all trees. Note that bagging for classification trees proceeds in a very similar fashion to bagging for regression trees. Of course, a different approach is taken to combine the predictions from the M bootstrap samples; this can be done, e.g., via majority vote.

One downside to bagging is that the M trees can be very similar, i.e., they can be highly correlated. The random forests approach (Breiman, 2001a) is an extension of bagging that decorrelates the M trees. For each tree, rather than all predictor variables being available for each split, a random sample of $\mathcal{M}$ is taken at each split. In fact, a different random sample of $\mathcal{M}$ predictor variables is considered at each split, for each tree. The result is that the M trees tend to be far more different than for bagging. Sometimes, guidelines such as $\mathcal{M} \approx \sqrt{p}$ are used to select $\mathcal{M}$, where p is the total number of predictor variables. We demonstrate a different approach in Section 7.4.2. Note that bagging is a special case of random forests, i.e., with $\mathcal{M} = p$, and so it will not be considered separately hereafter. Further material on random forests is deferred to Section 7.4.2, and we continue this chapter with gradient boosting.

5.7 GRADIENT BOOSTING

5.7.1 Overview

Boosting is a general approach to supervised learning, that generates an ensemble with M members from the training set. The ensemble members are generated sequentially, where the current one is created from a base learning algorithm using the residuals from the previous member as the response variable. We will restrict our discussion to decision tree learners. Unlike random forests, boosting does not use bootstrap sampling. It uses the entire dataset, or some subsample thereof, to generate the ensemble. There follows some pseudocode for a straightforward boosting algorithm.

A Straightforward Boosting Algorithm

read in training data $\mathbf{X}$, response Y, M, d, ω
initialize $f(\mathbf{x}) = 0$, $e_0 = y_0$
set number of splits per tree to d
for $m = 1, 2, \ldots, M$
    fit tree $f_m(\mathbf{x})$ to $(\mathbf{X}, e_{m-1})$
    $f(\mathbf{x}) = f(\mathbf{x}) + \omega f_m(\mathbf{x})$
    $e_m(\mathbf{x}) = e_{m-1}(\mathbf{x}) - \omega f_m(\mathbf{x})$
end for
return $f(\mathbf{x})$

The reader is encouraged to reflect on the relationship between this simple pseudocode and (5.1) and (5.2). In the terminology used in Section 5.1, we can think of final error as

$$e_M(\mathbf{x}) \equiv \mathcal{E}_{\text{training}}.$$

Note that ω is a parameter that controls the learning rate, i.e., a small number, and d is the tree depth. Of course, there is a tradeoff between the respective values of ω, d, and M.

Towards the end of Section 5.1, we concluded that we need to move away from thinking only about an error and consider an error together with some way to prevent or mitigate overfitting. The role of ω in the simple boosting algorithm already described is precisely to reduce overfitting. The `XGBoost.jl` package in Julia is an implementation of extreme gradient boosting (XGBoost), which is based on the idea of a gradient boosting machine (Friedman, 2001). A nice description of XGBoost is given in Chen and Guestrin (2016), where the starting point is a high-level description of the learning problem as the minimization of

$$\text{objective function} = \text{loss function} + \text{regularization term}.$$

In the very simple boosting algorithm described above, the regularization term is not present (Chen and Guestrin, 2016). It is an innovative feature of the XGBoost model which penalizes the model complexity beyond the traditional shrinkage parameter represented above by ω and can greatly improve predictive performance. Given the added complexity, the clearest way to explain XGBoost is by example, as follows.

The first code block has functions used to do 5-fold cross-validation for the boosting learners. The function `tunegrid_xgb()` creates a grid of the learning rate η and tree depth values d that the classification learners will be trained on. Including values for the number of boosting iterations (rounds) M performed could be a valuable addition to the tuning grid. We set this parameter value to 1000 for ease of exposition. The parameter η is a specific implementation of ω described above. Users of R may be familiar with the `gbm` package (Ridgeway, 2017), which also performs gradient boosting. The `shrinkage` parameter of the `gbm()` function from this package is another implementation of ω. The function `binomial_dev()` calculates the binomial deviance (5.8), which is used to train the learners, and `binary_mcr()` calculates the misclassification rate for the labeled outcomes. Note that we are using the `MLBase.jl` infrastructure to do the cross-validation of our boosting learners. It contains a general set of functions to train any supervised learning model using a variety of re-sampling schemes and evaluation metrics. The `cvERR_xgb()` function is used by `cross_validate()` from `MLBase.jl` to return the error metric of choice for each re-sample iteration being done. The `cvResult_xgb()` is the master control function for our cross-validation scheme, it uses the `MLBase.jl` functions `cross_validate()` and `StratifiedKfold()` to do stratified cross-validation on the outcome. The cross-validation indices are passed to `XGBoost` so it can make the predictions and the error metric results are summarized and returned as an array.

```julia
## CV functions for classification
using Statistics, StatsBase

## Array of parameters to train models on
## Includes two columns for the error mean and sd
function tunegrid_xgb(eta, maxdepth)
    n_eta = size(eta)[1]
    md = collect(1:maxdepth)
    n_md = size(md)[1]

    ## 2 cols to store the CV results
    res = zeros(*(maxdepth, n_eta), 4)
    n_res = size(res)[1]

    ## populate the res matrix with the tuning parameters
    res[:,1] = repeat(md, inner = Int64(/(n_res, n_md)))
    res[:,2] = repeat(eta, outer = Int64(/(n_res, n_eta)))

    return(res)
end

## MCR
function binary_mcr(yp, ya; trace = false)
    yl = Array{Int8}(yp .> 0.5)
    disagree = Array{Int8}(ya .!= yl)
    mcr = mean(disagree)
    if trace
        #print(countmap(yl))
        println("yl: ", yl[1:4])
        println("ya: ", ya[1:4])
        println("disagree: ", disagree[1:4])
        println("mcr: ", mcr)
    end
    return( mcr )
end

## Used by binomial_dev
## works element wise on vectors
function bd_helper(yp, ya)
    e_const = 1e-16
    pneg_const = 1.0 - yp
    if yp < e_const
        res = +(*(ya, log(e_const)), *( -(1.0, ya), log(-(1.0, e_const))))
    elseif pneg_const < e_const
        res = +(*(ya, log(-(1.0, e_const))), *( -(1.0, ya), log(e_const)))
    else
        res = +(*(ya, log(yp)), *( -(1.0, ya), log(pneg_const)))
    end
    return res
end
```

```
## Binomial Deviance
function binomial_dev(yp, ya) #::Vector{T}) where {T <: Number}
    res = map(bd_helper, yp, ya)
    dev = *(-2, sum(res))
    return(dev)
end

## functions used with MLBase.jl
function cvERR_xgb(model, Xtst, Ytst, error_fun; trace = false)
    y_p = XGBoost.predict(model, Xtst)
    if(trace)
        println("cvERR: y_p[1:5] ", y_p[1:5])
        println("cvERR: Ytst[1:5] ", Ytst[1:5])
        println("cvERR: error_fun(y_p, Ytst) ", error_fun(y_p, Ytst))
    end
    return(error_fun(y_p, Ytst))
end

function cvResult_xgb(Xmat, Yvec, cvparam, tunegrid; k = 5, n = 50,
    trace = false )

    result = deepcopy(tunegrid) ## tunegrid could have two columns on input
    n_tg = size(result)[1]

    ## k-Fold CV for each combination of parameters
    for i = 1:n_tg
        scores = cross_validate(
            trind -> xgboost(Xmat[trind,:], 1000, label = Yvec[trind],
                param = cvparam, max_depth = Int(tunegrid[i,1]),
                    eta = tunegrid[i,2]),
            (c, trind) -> cvERR_xgb(c, Xmat[trind,:], Yvec[trind],
                binomial_dev, trace = true),
                ## total number of samples
            n,
                ## Stratified CV on the outcome
            StratifiedKfold(Yvec, k)
        )
        if(trace)
            println("cvResult_xgb: scores: ", scores)
            println("size(scores): ", size(scores))
        end
        result[i, 3] = mean(scores)
        result[i, 4] = std(scores)
    end

    return(result)
end
```

5.7.2 Beer Data

The next code block illustrates how the functions detailed in the previous code block can be used to classify beers in the beer data as either lagers or ales. We start by creating an 80-20 training test split of the data, stratified on outcome. Next, there is an array of parameters that we will pass to all of our XGBoost learners. Besides the parameters being used to generate the tuning grid, the others will be held constant across learners. The `tunegrid_xgb()` function is used to generate a training grid for η and d. Increasing the tree depth and η is one way to increase the complexity of the XGBoost learner. The `cvResult_xgb()` function is used to generate the training results illustrated in Figure 5.8, which suggest that learners with small η values and medium sized trees are classifying the beers best.

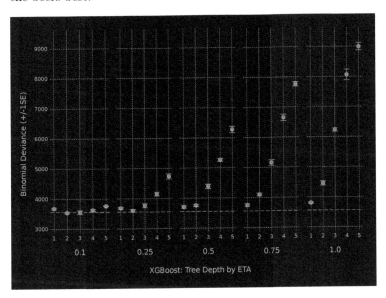

Figure 5.8 Summary of cross-validation performance for XG-Boost from 5-fold cross-validation.

The horizontal line in Figure 5.8 is one standard error above the smallest cross-validation training deviance. Because there are a number of learners with error bars overlapping the line, we use the one-standard error method for choosing the final learner. This method was first proposed by Breiman et al. (1984) in the context of classification and regression trees and is discussed further by Hastie et al. (2009). The method finds the simplest, i.e., most regularized, learner configuration within one standard error above the numerically best learner, i.e., the configuration with the lowest error rate. The results are depicted in Figure 5.9. We choose the learner with $\eta = 0.1$ and a tree depth of 2. This learner is run on the test set data and has a misclassification rate of 8.2% and a deviance of 4458.08. As expected, the deviance is larger than the value found by cross-validation but still very reasonable.

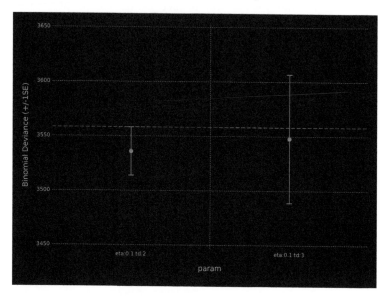

Figure 5.9 Summary of results for the 1 SE method applied to the XGBoost analysis of the beer data.

```
## 80/20 split
splt = 0.8
y_1_ind = N_array[df_recipe[:y] .== 1]
y_0_ind = N_array[df_recipe[:y] .== 0]

tr_index = sort(
  vcat(
    sample(y_1_ind, Int64(floor(*(N_y_1, splt))), replace = false),
    sample(y_0_ind, Int64(floor(*(N_y_0, splt))), replace = false)
  )
)
tst_index = setdiff(N_array, tr_index)

df_train = df_recipe[tr_index, :]
df_test = df_recipe[tst_index, :]

## Check proportions
println("train: n y: $(sum(df_train[:y]))")
println("train: % y: $(mean(df_train[:y]))")
println("test: n y: $(sum(df_test[:y]))")
println("test: % y: $(mean(df_test[:y]))")

## Static boosting parameters
```

```
param_cv = [
  "silent" => 1,
  "verbose" => 0,
  "booster" => "gbtree",
  "objective" => "binary:logistic",
  "save_period" => 0,
  "subsample" => 0.75,
  "colsample_bytree" => 0.75,
  "alpha" => 0,
  "lambda" => 1,
  "gamma" => 0
 ]

## training data: predictors are sparse matrix
tmp = convert(Array, df_train[setdiff(names(df_train), [:y])] )
xmat_tr = convert(SparseMatrixCSC{Float64,Int64},
                  collect(Missings.replace(tmp, 0)))

## CV Results

## eta shrinks the feature weights to help prevent overfilling by making
## the boosting process more conservative
## max_depth is the maximum depth of a tree; increasing this value makes
## the boosting process more complex
tunegrid = tunegrid_xgb([0.1, 0.25, 0.5, 0.75, 1], 5)
N_tr = size(xmat_tr)[1]

cv_res = cvResult_xgb(xmat_tr, y_tr, param_cv, tunegrid,
  k = 5, n = N_tr, trace = true )

## dataframe for plotting
cv_df = DataFrame(tree_depth = cv_res[:, 1],
  eta = cv_res[:, 2],
  mean_error = cv_res[:, 3],
  sd_error = cv_res[:, 4],
)
cv_df[:se_error] = map(x -> x / sqrt(5), cv_df[:sd_error])
cv_df[:mean_min] = map(-, cv_df[:mean_error], cv_df[:se_error])
cv_df[:mean_max] = map(+, cv_df[:mean_error], cv_df[:se_error])

## configurations within 1 se
min_dev = minimum(cv_df[:mean_error])
min_err = filter(row -> row[:mean_error] == min_dev, cv_df)
one_se = min_err[1, :mean_max]
possible_models = filter(row -> row[:mean_error] <= one_se, cv_df)

#####
## Test Set Predictions

tst_results = DataFrame(
 eta = Float64[],
 tree_depth = Int64[],
 mcr = Float64[],
 dev = Float64[])

## using model configuration selected above
pred_tmp = XGBoost.predict(xgboost(xmat_tr, 1000, label = y_tr,
  param = param_cv, eta =0.1 , max_depth = 2), xmat_tst)
tmp = [0.1, 2, binary_mcr(pred_tmp, y_tst), binomial_dev(pred_tmp, y_tst)]
push!(tst_results, tmp)

#####
```

```
## Variable importance from the overall data
fit_oa = xgboost(xmat_oa, 1000, label = y_oa, param = param_cv, eta = 0.1,
   max_depth = 2)

## names of features
names_oa =  map(string, setdiff(names(df_recipe), [:y]))
fit_oa_imp = importance(fit_oa, names_oa)

## DF for plotting
N_fi = length(fit_oa_imp)

imp_df = DataFrame(fname = String[], gain = Float64[], cover = Float64[])

for i = 1:N_fi
    tmp = [fit_oa_imp[i].fname, fit_oa_imp[i].gain, fit_oa_imp[i].cover]
    push!(imp_df, tmp)
end

sort!(imp_df, :gain, rev=true)
```

The learner chosen by the one-standard error method is run
on the full data to determine which variables are most predictive
of the beer types. We use the full dataset because the important
variables could differ from those found in the training set. The
results are displayed graphically in Figure 5.10.

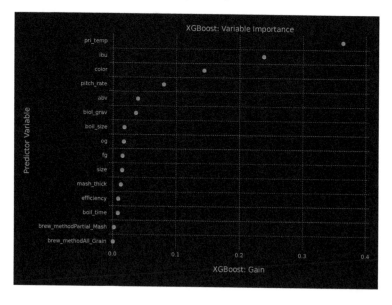

Figure 5.10 Variable importance plot from the XGBoost analysis of the beer data.

XGBoost measures variable importance in a number of ways, we will focus on the `gain` measure. It is the mean accuracy improvement brought on by creating a split in a tree on this variable across the boosting ensemble. The `gain` is scaled to be between 0 and 1. There are three important predictor variables for the beer data: primary brewing temperature; IBU; and colour units. This makes sense considering that it is well-known that ales are brewed at higher temperatures and tend have a higher bitterness rating (Oliver and Colicchio, 2011). Ales include many dark-coloured beers, which might account for the predictive ability of the colour variable.

5.7.3 Food Data

The following code block details how we went about training XGBoost regression learners on the `food` data. We have many of the same functions but with slightly different configurations. The `tunegrid_xgb_reg()` function adds a third parameter to the tuning grid, but it is otherwise the same as the one we used for the classification learners. The XGBoost models are run for 1000 iterations. The error metrics have changed: now, they are the MAE, calculated using `medae()`, and the RMSE, calculated using `rmse()`. Note that `cvResult_xgb_reg()` plays an analogous role to `cvResult_xgb()`, controlling the cross-validation process. Specifically, it carries out unstratified cross-validation and trains the learners using the `medae()` function.

```
## CV Functions for Regression

function tunegrid_xgb_reg(eta, maxdepth, alpha)
    n_eta = size(eta)[1]
    md = collect(1:maxdepth)
    n_md = size(md)[1]
    n_a = size(alpha)[1]

    ## 2 cols to store the CV results
    res = zeros(*(maxdepth, n_eta, n_a), 5)
    n_res = size(res)[1]
    println("N:", n_res, " e: ", n_eta, " m: ", n_md, " a: ", n_a)
    ## populate the res matrix with the tuning parameters
    n_md_i = Int64(/(n_res, n_md))
    res[:,1] = repeat(md, outer=n_md_i)
    res[:,2] = repeat(eta, inner = Int64(/(n_res, n_eta)))
    res[:,3] = repeat(repeach(alpha, (n_md)), outer=n_a)

    return(res)
end
```

```
function medae(yp, ya)
    ## element wise operations
    res = map(yp, ya) do x,y
        dif = -(x, y)
        return(abs(dif))
    end
    return(median(res))
end

function rmse(yp, ya)
    ## element wise operations
    res = map(yp, ya) do x,y
        dif = -(x, y)
        return(dif^2)
    end
    return(sqrt(mean(res)))
end

function cvResult_xgb_reg(Xmat, Yvec, cvparam, tunegrid; k = 5, n = 50,
    trace = false )

    result = deepcopy(tunegrid)
    n_tg = size(result)[1]

    ## k-Fold CV for each combination of parameters
    for i = 1:n_tg
        scores = cross_validate(
            ## num_round investigate
            trind -> xgboost(Xmat[trind,:], 1000, label = Yvec[trind],
                param = cvparam, max_depth = Int(tunegrid[i,1]),
                    eta = tunegrid[i,2], alpha = tunegrid[i,3]),
            (c, trind) -> cvERR_xgb(c, Xmat[trind,:], Yvec[trind], medae,
                trace = true),
            n, ## total number of samples
            Kfold(n, k)
        )
        if(trace)
            println("cvResult_xgb_reg: scores: ", scores)
            println("size(scores): ", size(scores))
        end
        result[i, 4] = mean(scores)
        result[i, 5] = std(scores)
    end

    return(result)
end
```

The following code block illustrates how the functions defined above can be used to train our XGBoost regression learners. Start by making a 80-20 training-test split of the data. Then define the static boosting parameters for the XGBoost learners. The tuning grid is generated for values of η, d, and α, an L1 regularization term on the learner weights. The α parameter is one of XGBoost's regularization terms and, when included in the model tuning, can produce very predictive and fast models when there are a large number of predictor variables. The 5-fold cross-validation results are stored in the `cv_res_reg` array and visualized in Figure 5.11. Once again, there are many candidate learners to choose from. Learners with larger tree depths and smaller values of η and α seem to perform best on the training set.

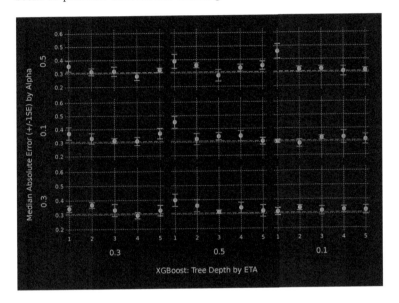

Figure 5.11 Summary of results for 5-fold cross-validation from the XGBoost analysis of the `food` data.

```
## 80/20 train test split
splt = 0.8

tr_index = sample(N_array, Int64(floor(*(N, splt))), replace = false)
tst_index = setdiff(N_array, tr_index)

df_train = df_food[tr_index, :]
df_test = df_food[tst_index, :]

## train data: predictors are sparse matrix
y_tr = convert(Array{Float64}, df_train[:gpa])
tmp = convert(Array, df_train[setdiff(names(df_train), [:gpa])] )
xmat_tr = convert(SparseMatrixCSC{Float64,Int64},
  collect(Missings.replace(tmp, 0)))

## Static boosting parameters
param_cv_reg = [
  "silent" => 1,
  "verbose" => 0,
  "booster" => "gbtree",
  "objective" => "reg:linear",
```

```
 "save_period" => 0,
 "subsample" => 0.75,
 "colsample_bytree" => 0.75,
 "lambda" => 1,
 "gamma" => 0
]

## CV Results

## eta shrinks the feature weights to help prevent overfilling by making
## the boosting process more conservative
## max_depth is the maximum depth of a tree; increasing this value makes
## the boosting process more complex
## alpha is an L1 regularization term on the weights; increasing this value
## makes the boosting process more conservative
tunegrid = tunegrid_xgb_reg([0.1, 0.3, 0.5], 5, [0.1, 0.3, 0.5])
N_tr = size(xmat_tr)[1]

cv_res_reg = cvResult_xgb_reg(xmat_tr, y_tr, param_cv_reg, tunegrid,
 k = 5, n = N_tr, trace = true )

## add the standard error
cv_res_reg = hcat(cv_res_reg, ./(cv_res_reg[:,5], sqrt(5)))

## dataframe for plotting
cv_df_r = DataFrame(tree_depth = cv_res_reg[:, 1],
  eta = cv_res_reg[:, 2],
  alpha = cv_res_reg[:, 3],
  medae = cv_res_reg[:, 4],
  medae_sd = cv_res_reg[:, 5]
)
cv_df_r[:medae_se] = map(x -> /(x , sqrt(5)), cv_df_r[:medae_sd])
cv_df_r[:medae_min] = map(-, cv_df_r[:medae], cv_df_r[:medae_se])
cv_df_r[:medae_max] = map(+, cv_df_r[:medae], cv_df_r[:medae_se])

min_medae = minimum(cv_df_r[:medae])
min_err = filter(row -> row[:medae] == min_medae, cv_df_r)
one_se = min_err[1, :medae_max]

#######
## Test Set Predictions

tst_results = DataFrame(
  eta = Float64[],
  alpha = Float64[],
  tree_depth = Int64[],
  medae = Float64[],
  rmse = Float64[])

## using configuration chosen above
pred_tmp = XGBoost.predict(xgboost(xmat_tr, 1000, label = y_tr,
  param = param_cv_reg, eta =0.1 , max_depth = 1, alpha = 0.1), xmat_tst)
tmp = [0.1, 0.1, 1, medae(pred_tmp, y_tst), rmse(pred_tmp, y_tst)]
push!(tst_results, tmp)

#####
## Variable Importance from the overall data

fit_oa = xgboost(xmat_oa, 1000, label = y_oa, param = param_cv_reg,
  eta = 0.1, max_depth = 1, alpha = 0.1)

## names of features
```

```
# names_oa = convert(Array{String,1}, names(df_food))
names_oa = map(string, setdiff(names(df_train), [:gpa]))
fit_oa_imp = importance(fit_oa, names_oa)

## DF for plotting
N_fi = length(fit_oa_imp)

imp_df = DataFrame(fname = String[], gain = Float64[], cover = Float64[])

for i = 1:N_fi
    tmp = [fit_oa_imp[i].fname, fit_oa_imp[i].gain, fit_oa_imp[i].cover]
    push!(imp_df, tmp)
end

sort!(imp_df, :gain, rev=true)
```

Given the plethora of candidate learners, we again apply the one-standard error method to select the best candidate learner parameter configuration. The results are summarized in Figure 5.12. There are six candidates to choose from. We opt for the stump learner with $\eta = \alpha = 0.1$ and $d = 1$, because it has the most regularization and the simplest tree structure. When run on the test set, the MAE is 0.533 and the RMSE is 0.634.

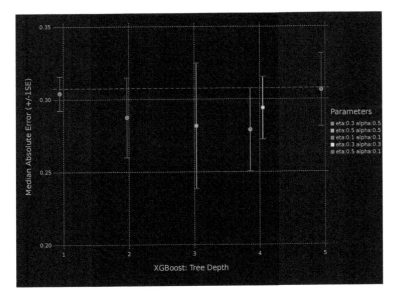

Figure 5.12 Summary of results for the 1 SE method applied to the XGBoost analysis of the food data.

This learner was used to determine the variable importance on the overall data. The results are displayed in Figure 5.13. The left-hand panel contains a violin plot of the overall gain scores. It is obvious that most of the predictor variables have little-to-no effect on the prediction of student GPA. There are many weak predictor variables that play a part in successfully predicting student GPA from these data. Based on the variable importance plot in Figure 5.13, the important predictor variables in the learner reflect: a highly educated father; a poorly educated mother; being able to estimate the calories in food; and measuring one's body weight.

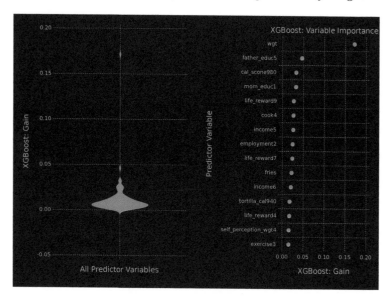

Figure 5.13 Violin plot and variable importance plot from the XGBoost analysis of the food data.

5.8 COMMENTS

Selected supervised learning techniques have been covered, along with Julia code needed for implementation. The primary goal of this book is to expose the reader to Julia and, as mentioned in the Preface, it is not intended as a thorough introduction to data science. Certainly, this chapter cannot be considered a thorough introduction to supervised learning. Some of the more notable absentees include generalized additive models, neural networks, and their many extensions as well as assessment tools such as sensitivity, specificity, and receiver operating characteristic (ROC) curves. Random forests were discussed in this chapter; however, their implementation and demonstration is held over for Section 7.4.2.

The idea that some learners are more prone to overfitting than others was touched upon in Section 5.1. This is a topic that deserves careful thought. For example, we applied the one standard error rule to reduce the complexity when fitting our kNN learner, by choosing the smallest k when several values gave similar performance. The reader may be interested to contrast the use of gradient boosting in Section 5.7 with that of random forests (Section 7.4.2). It is, perhaps, difficult to escape the feeling that an explicit penalty for overfitting might benefit random forests and some other supervised learning approaches that do not typically use one. When using random forests for supervised learning, it is well known that increasing the size of the forest does not lead to over-fitting (Hastie et al., 2009, Chp. 15). When using it for regression, utilizing unpruned trees in the forest can lead to over-fitting which will increase the variance of the predictions (Hastie et al., 2009, Chp. 15). Despite this, few software packages provide this parameter for performance tuning.

## CHAPTER 6

Unsupervised Learning

U NSUPERVISED LEARNING includes a host of exploratory techniques that can be used to gain insight into data, often when there is no outcome variable or clear target. Although around for over a century, principal components analysis remains a data science staple and it is introduced along with `MultivariateStats.jl`. Probabilistic principal components analysis (PPCA) is considered next and an expectation-maximization algorithm for PPCA is implemented in Julia — this is the first of two examples in this chapter that illustrate Julia as a base language for data science. Clustering, or unsupervised classification, is introduced along with the famous k-means clustering technique and `Clustering.jl`. A mixture of probabilistic principal components analyzers is then discussed and implemented, thereby providing the second example of Julia as a base language for data science.

6.1 INTRODUCTION

Unsupervised learning differs from supervised learning (Chapter 5) in that there are no labelled observations. It may be that there are labels but none are known *a priori*, it may be that some labels are known but the data are being treated as if unlabelled, or it may be the case that there are no labels *per se*. Because there are no labelled observations, it might be advantageous to lean towards statistical approaches and models — as opposed to machine learning — when tackling unsupervised learning problems. In a statistical framework, we can express our uncertainty in mathematical terms, and use clearly defined models and procedures with well-understood assumptions. Statistical approaches have many practical advantages, e.g., we may be able to write down likelihoods and compare models using likelihood-based criteria.

Suppose n realizations $\mathbf{x}_1, \ldots, \mathbf{x}_n$ of p-dimensional random variables $\mathbf{X}_1, \ldots, \mathbf{X}_n$ are observed, where $\mathbf{X}_i = (X_{i1}, X_{i2}, \ldots, X_{ip})'$ for $i = 1, \ldots, n$. As mentioned in Section 3.1, the quantity $\mathscr{X} = (\mathbf{X}_1, \mathbf{X}_2, \ldots, \mathbf{X}_n)'$ can be considered a random matrix and its realization is called a data matrix. Recall also that a matrix $\mathbf{A}$ with all entries constant is called a constant matrix. Now, each $\mathbf{X}_i$ is a $p \times 1$ random vector and, for our present purposes, it will be helpful to consider some results for random vectors. Let $\mathbf{X}$ and $\mathbf{Y}$ be $p \times 1$ random vectors. Then,

$$\mathbb{E}[\mathbf{X} + \mathbf{Y}] = \mathbb{E}[\mathbf{X}] + \mathbb{E}[\mathbf{Y}],$$

and, if $\mathbf{A}$, $\mathbf{B}$, and $\mathbf{C}$ are $m \times n$, $p \times q$, and $m \times q$ constant matrices, respectively, then

$$\mathbb{E}[\mathbf{A}\mathbf{X}\mathbf{B} + \mathbf{C}] = \mathbf{A}\mathbb{E}[\mathbf{X}]\mathbf{B} + \mathbf{C}.$$

Suppose $\mathbf{X}$ has mean $\boldsymbol{\mu}$. Then, the covariance matrix of $\mathbf{X}$ is

$$\boldsymbol{\Sigma} = \mathbb{V}\mathrm{ar}[\mathbf{X}] = \mathbb{E}[(\mathbf{X} - \boldsymbol{\mu})(\mathbf{X} - \boldsymbol{\mu})'].$$

Suppose p-dimensional $\mathbf{X}$ has mean $\boldsymbol{\mu}$ and q-dimensional $\mathbf{Y}$ has mean $\boldsymbol{\theta}$. Then,

$$\mathrm{Cov}[\mathbf{X}, \mathbf{Y}] = \mathbb{E}[(\mathbf{X} - \boldsymbol{\mu})(\mathbf{Y} - \boldsymbol{\theta})'].$$

The covariance matrix $\boldsymbol{\Sigma}$ of p-dimensional $\mathbf{X}$ can be written

$$\boldsymbol{\Sigma} = \begin{pmatrix} \sigma_{11} & \sigma_{12} & \cdots & \sigma_{1p} \\ \sigma_{21} & \sigma_{22} & \cdots & \sigma_{2p} \\ \vdots & \vdots & \ddots & \vdots \\ \sigma_{p1} & \sigma_{p2} & \cdots & \sigma_{pp} \end{pmatrix}.$$

The elements σ_{ii} are variances and σ_{ij}, where $i \neq j$, are the covariances. The matrix $\boldsymbol{\Sigma}$ is symmetric because $\sigma_{ij} = \sigma_{ji}$. Also, $\boldsymbol{\Sigma}$ is positive semi-definite, i.e., for any $p \times 1$ constant vector $\mathbf{a} = (a_1, \ldots, a_p)'$,

$$\mathbf{a}'\boldsymbol{\Sigma}\mathbf{a} \geq 0.$$

Suppose $\mathbf{C}$ is a $q \times p$ constant matrix and $\mathbf{c}$ is a $q \times 1$ constant vector. Then, the covariance matrix of $\mathbf{Y} = \mathbf{C}\mathbf{X} + \mathbf{c}$ is

$$\mathbb{V}\mathrm{ar}[\mathbf{Y}] = \mathbf{C}\boldsymbol{\Sigma}\mathbf{C}'.$$

In general, the covariance matrix $\boldsymbol{\Sigma}$ is positive semi-definite. Therefore, the eigenvalues of $\boldsymbol{\Sigma}$ are non-negative, and are denoted

$$\kappa_1 \geq \kappa_2 \geq \cdots \geq \kappa_p \geq 0.$$

Eigenvalues are also known as characteristic roots. Write $\mathbf{v}_i$ to denote the eigenvector corresponding to the eigenvalue κ_i, for $i = 1, \ldots, p$. Eigenvectors are also known as characteristic vectors. Without loss of generality, it can be assumed that the eigenvectors are orthonormal, i.e.,

$$\mathbf{v}_i'\mathbf{v}_i = 1$$

and

$$\mathbf{v}_i'\mathbf{v}_j = 0$$

for $i \neq j$. The eigenvalues and eigenvectors satisfy

$$\boldsymbol{\Sigma}\mathbf{v}_i = \kappa_i\mathbf{v}_i, \tag{6.1}$$

for $i = 1, \ldots, p$.

Write $\mathbf{V} = (\mathbf{v}_1, \ldots, \mathbf{v}_p)$ and

$$\mathbf{K} = \mathrm{diag}\{\kappa_1, \ldots, \kappa_p\} = \begin{pmatrix} \kappa_1 & 0 & \cdots & 0 \\ 0 & \kappa_2 & \cdots & 0 \\ \vdots & \vdots & \ddots & \vdots \\ 0 & 0 & \cdots & \kappa_p \end{pmatrix}.$$

Then, (6.1) can be written

$$\boldsymbol{\Sigma}\mathbf{V} = \mathbf{V}\mathbf{K}.$$

Note that $\mathbf{V}'\mathbf{V} = \mathbf{I}_p$, and

$$\boldsymbol{\Sigma} = \mathbf{V}\mathbf{K}\mathbf{V}' = \mathbf{V}\mathbf{K}^{1/2}\mathbf{V}'\mathbf{V}\mathbf{K}^{1/2}\mathbf{V}',$$

where

$$\mathbf{K}^{1/2} = \mathrm{diag}\{\sqrt{\kappa_1}, \ldots, \sqrt{\kappa_p}\}.$$

The determinant of the covariance matrix $\boldsymbol{\Sigma}$ can be written

$$|\boldsymbol{\Sigma}| = |\mathbf{V}\mathbf{K}\mathbf{V}'| = |\mathbf{K}|$$

$$= \kappa_1 \times \kappa_2 \times \cdots \times \kappa_p = \prod_{i=1}^{p} \kappa_i.$$

The total variance of the covariance matrix $\mathbf{\Sigma}$ can be written

$$\text{tr}\{\mathbf{\Sigma}\} = \text{tr}\{\mathbf{K}\} = \kappa_1 + \kappa_2 + \cdots + \kappa_p.$$

The determinant and total variance are sometimes used as summaries of the total scatter amongst the p variables. Note that some background on eigenvalues is provided in Appendix C.1.

6.2 PRINCIPAL COMPONENTS ANALYSIS

In general, observed data will contain correlated variables. The objective of principal components analysis is to replace the observed variables by a number of uncorrelated variables that explain a sufficiently large amount of the variance in the data. Before going into mathematical details, it will be helpful to understand the practical meaning of a principal component. The first principal component is the direction of most variation in the data. The second principal component is the direction of most variation in the data conditional on it being orthogonal to the first principal component. In general, for $r \in (1, p]$, the rth principal component is the direction of most variation in the data conditional on it being orthogonal to the first $r - 1$ principal components. There are several mathematical approaches to motivating principal components analysis and the approach followed here is based on that used by Fujikoshi et al. (2010).

Let $\mathbf{X}$ be a p-dimensional random vector with mean $\boldsymbol{\mu}$ and covariance matrix $\mathbf{\Sigma}$. Let $\kappa_1 \geq \kappa_2 \geq \cdots \geq \kappa_p \geq 0$ be the (ordered) eigenvalues of $\mathbf{\Sigma}$, and let $\mathbf{v}_1, \mathbf{v}_2, \ldots, \mathbf{v}_p$ be the corresponding eigenvectors. The ith principal component of $\mathbf{X}$ is

$$W_i = \mathbf{v}_i'(\mathbf{X} - \boldsymbol{\mu}), \tag{6.2}$$

for $i = 1, \ldots, p$. Which we can write as

$$\mathbf{W} = \mathbf{V}'(\mathbf{X} - \boldsymbol{\mu}), \tag{6.3}$$

where $\mathbf{W} = (W_1, \ldots, W_p)$ and $\mathbf{V} = (\mathbf{v}_1, \ldots, \mathbf{v}_p)$.

Now, it is helpful to consider some properties of $\mathbf{W}$. First of all, because $\mathbb{E}[\mathbf{X}] = \boldsymbol{\mu}$,

$$\mathbb{E}[\mathbf{W}] = \mathbf{0}.$$

Recalling that $\mathbb{V}\text{ar}[\mathbf{X}] = \mathbf{\Sigma} = \mathbf{VKV}'$, we have

$$\begin{aligned} \mathbb{V}\text{ar}[\mathbf{W}] &= \mathbf{V}'\mathbb{V}\text{ar}[\mathbf{X}]\mathbf{V} \\ &= \mathbf{V}'(\mathbf{VKV}')\mathbf{V} \\ &= (\mathbf{V}'\mathbf{V})\mathbf{K}(\mathbf{V}'\mathbf{V}) = \mathbf{K}. \end{aligned} \tag{6.4}$$

Rearranging (6.3),
$$\mathbf{V}'\mathbf{X} = \mathbf{V}'\boldsymbol{\mu} + \mathbf{W},$$

and left-multiplying by $\mathbf{V}$ gives

$$\mathbf{X} = \boldsymbol{\mu} + \mathbf{V}\mathbf{W} = \boldsymbol{\mu} + \sum_{i=1}^{p} \mathbf{v}_i W_i.$$

From (6.4), we can develop the result that

$$\sum_{i=1}^{p} \mathbb{Var}[W_i] = \mathrm{tr}\{\mathbf{K}\} = \mathrm{tr}\{\mathbf{K}\mathbf{V}'\mathbf{V}\}$$
$$= \mathrm{tr}\{\mathbf{V}\mathbf{K}\mathbf{V}'\} = \mathrm{tr}\{\boldsymbol{\Sigma}\}.$$

Now, the proportion of the total variation explained by the jth principal component is

$$\frac{\kappa_i}{\mathrm{tr}\{\mathbf{K}\}} = \frac{\kappa_i}{\mathrm{tr}\{\boldsymbol{\Sigma}\}}.$$

Therefore, the proportion of the total variation explained by the first k principal components is

$$\frac{\sum_{i=1}^{k} \kappa_i}{\mathrm{tr}\{\boldsymbol{\Sigma}\}}.$$

To understand how the theoretical explanation of a principal component relates to the intuition given at the beginning of this section, consider
$$W = \mathbf{v}'(\mathbf{X} - \boldsymbol{\mu})$$

such that $\mathbf{v}'\mathbf{v} = 1$. Now,

$$\mathbb{Var}[W] = \mathbf{v}'\boldsymbol{\Sigma}\mathbf{v}$$

and we can write

$$\mathbf{v} = a_1\mathbf{v}_1 + \cdots + a_p\mathbf{v}_p = \mathbf{V}\mathbf{a}$$

for $\mathbf{a}'\mathbf{a} = 1$. Therefore,

$$\mathbb{Var}[W] = a_1^2\kappa_1 + \cdots + a_p^2\kappa_p. \tag{6.5}$$

Now, maximizing (6.5) subject to the constraint $\mathbf{a}'\mathbf{a} = 1$ gives $\mathbb{Var}[W] = \kappa_1$, i.e., $a_1 = 1$ and $a_j = 0$ for $j > 1$. In other words,

$\mathbb{V}\text{ar}[W]$ is maximized at $W = W_1$, i.e., the first principal component.

Now suppose that W is uncorrelated with the first $r - 1 < p$ principal components $W_1, \ldots, W_{r-1}$. We have

$$\text{Cov}[W, W_i] = \mathbb{E}[\mathbf{v}'(\mathbf{X} - \boldsymbol{\mu})(\mathbf{X} - \boldsymbol{\mu})'\mathbf{v}_i] = \kappa_i c_i$$

and so $c_i = 0$ for $i = 1, \ldots, r - 1$. Maximizing (6.5) subject to the constraint $c_i = 0$, for $i = 1, \ldots, r - 1$, gives $\mathbb{V}\text{ar}[W] = \kappa_r$. In other words, $\mathbb{V}\text{ar}[W]$ is maximized at $W = W_r$, i.e., the rth principal component.

Illustrative Example: Crabs Data

The following code block performs principal components analyses on the **crabs** data.

```julia
using LinearAlgebra

## function to do pca via SVD
function pca_svd(X)
    n,p = size(X)
    k = min(n,p)
    S = svd(X)
    D = S.S[1:k]
    V = transpose(S.Vt)[:,1:k]
    sD = /(D, sqrt(n-1))
    rotation = V
    projection = *(X, V)
    return(Dict(
      "sd" => sD,
      "rotation" => rotation,
      "projection" => projection
    ))
end

## crab_mat_c is a Float array with  5 continuous mearures
## each variable is mean centered
crab_mat = convert(Array{Float64}, df_crabs[[:FL, :RW, :CL, :CW, :BD]])
mean_crab = mean(crab_mat, dims = 1)
crab_mat_c = crab_mat .- mean_crab

pca1 = pca_svd(crab_mat)

## df for plotting
## label is the combination of sp and sex
pca_df2 = DataFrame(
  pc1 = pca1["projection"][:,1],
  pc2 = pca1["projection"][:,2],
  pc3 = pca1["projection"][:,3],
  pc4 = pca1["projection"][:,4],
  pc5 = pca1["projection"][:,5],
  label = map(string, repeat(1:4, inner = 50))
)
```

More than 98% of the variation in the data is explained by the first principal component, with subsequent principal components accounting for relatively little variation (Table 6.1).

Table 6.1 The proportion, and cumulative proportion, of variation in the crabs data that is explained by the principal components (PCs).

PC	Prop. of Variation Explained	Cumulative Prop.
1	0.9825	0.9825
2	0.0091	0.9915
3	0.0070	0.9985
4	0.0009	0.9995
5	0.0005	1.0000

Interestingly, even though the second and third principal components explain a total of 1.61% of the variation in the data, they give very good separation of the four classes in the crabs dataset; in fact, they give much better separation of the classes than the first two principal components (Figures 6.1 and 6.2). This example illustrates that principal components that explain the vast majority of the variation within data still might be missing important information about subpopulations or, perhaps unknown, classes.

6.3 PROBABILISTIC PRINCIPAL COMPONENTS ANALYSIS

Probabilistic principal components analysis (PPCA; Tipping and Bishop, 1999b) is a data reduction technique that replaces p observed variables by $q < p$ latent components. The method works well when the q latent components explain a satisfactory amount of the variability in the p observed variables. In some situations, we might even have substantial dimension reduction, i.e., $q \ll p$. There are interesting relationships between PPCA and principal components analysis and, indeed, between PPCA and factor analysis (Spearman, 1904, 1927) — these relationships will be discussed later. In what follows, the notation and approach are similar to those used by McNicholas (2016a).

Consider independent p-dimensional random variables $\mathbf{X}_1, \ldots, \mathbf{X}_n$. The PPCA model can be written

$$\mathbf{X}_i = \boldsymbol{\mu} + \boldsymbol{\Lambda}\mathbf{U}_i + \boldsymbol{\epsilon}_i, \tag{6.6}$$

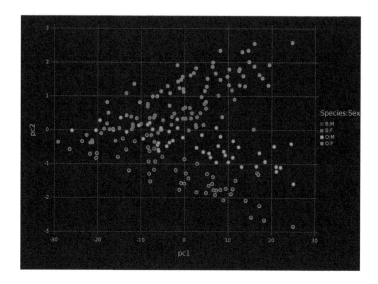

Figure 6.1 Scallerplot depicting the first two principal components for the **crabs** data, coloured by class.

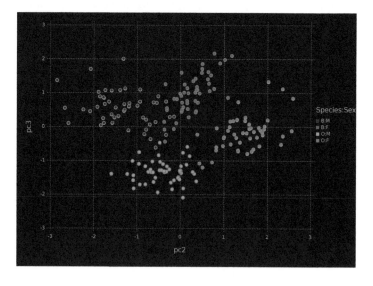

Figure 6.2 Scallerplot depicting the second and third principal components for the **crabs** data, coloured by class.

for $i = 1, \ldots, n$, where $\mathbf{\Lambda}$ is a $p \times q$ matrix of component (or factor) loadings, the latent component $\mathbf{U}_i \sim \mathcal{N}(\mathbf{0}, \mathbf{I}_q)$, and $\boldsymbol{\epsilon}_i \sim \mathcal{N}(\mathbf{0}, \psi \mathbf{I}_p)$, where $\psi \in \mathbb{R}^+$. Note that the $\mathbf{U}_i$ are independently distributed and independent of the $\boldsymbol{\epsilon}_i$, which are also independently distributed. From (6.6), it follows that the marginal distribution of $\mathbf{X}_i$ under the PPCA model is $\mathcal{N}(\boldsymbol{\mu}, \mathbf{\Lambda}\mathbf{\Lambda}' + \psi \mathbf{I}_p)$. There are

$$pq - \frac{1}{2}q(q-1) + 1$$

free parameters in the covariance matrix $\mathbf{\Lambda}\mathbf{\Lambda}' + \psi \mathbf{I}_p$ (Lawley and Maxwell, 1962). Therefore, the reduction in free covariance parameters under the PPCA model is

$$\frac{1}{2}p(p+1) - \left[pq - \frac{1}{2}q(q-1) + 1\right] = \frac{1}{2}\left[(p-q)^2 + p - q - 2\right],$$
(6.7)

and there is a reduction in the number of free parameters provided that (6.7) is positive, i.e., provided that

$$(p-q)^2 > q - p + 2.$$

The log-likelihood for p-dimensional $\mathbf{x}_1, \mathbf{x}_2, \ldots, \mathbf{x}_n$ from the PPCA model is

$$l(\boldsymbol{\mu}, \mathbf{\Lambda}, \mathbf{\Psi}) = \sum_{i=1}^{n} \log \phi(\mathbf{x}_i \mid \boldsymbol{\mu}, \mathbf{\Lambda}\mathbf{\Lambda}' + \psi \mathbf{I}_p)$$

$$= -\frac{np}{2}\log 2\pi - \frac{n}{2}\log|\mathbf{\Lambda}\mathbf{\Lambda}' + \psi \mathbf{I}_p| - \frac{n}{2}\operatorname{tr}\left\{\mathbf{S}(\mathbf{\Lambda}\mathbf{\Lambda}' + \psi \mathbf{I}_p)^{-1}\right\},$$
(6.8)

where

$$\mathbf{S} = \frac{1}{n}\sum_{i=1}^{n}(\mathbf{x}_i - \boldsymbol{\mu})(\mathbf{x}_i - \boldsymbol{\mu})'.$$
(6.9)

The maximum likelihood estimate for $\boldsymbol{\mu}$ is easily obtained by differentiating (6.8) with respect to $\boldsymbol{\mu}$ and setting the resulting score function equal to zero to get $\hat{\boldsymbol{\mu}} = \bar{\mathbf{x}}$. An EM algorithm can be used to obtain maximum likelihood estimates for $\mathbf{\Lambda}$ and $\mathbf{\Psi}$.

6.4   EM ALGORITHM FOR PPCA

6.4.1   Background: EM Algorithm

The expectation-maximization (EM) algorithm (Dempster et al., 1977) is an iterative procedure for finding maximum likelihood estimates when data are incomplete or treated as such. The EM algorithm iterates between an E-step and an M-step until some stopping rule is reached. On the E-step, the expected value of the complete-data log-likelihood is computed conditional on the current parameter estimates. On the M-step, this quantity is maximized with respect to the parameters to obtain (new) updates. Extensive details on the EM algorithm are provided by McLachlan and Krishnan (2008).

6.4.2 E-step

The complete-data comprise the observed $\mathbf{x}_1, \ldots, \mathbf{x}_n$ together with the latent components $\mathbf{u}_1, \ldots, \mathbf{u}_n$, where $\mathbf{u}_i = (u_{i1}, \ldots, u_{iq})'$. Now, noting that $\mathbf{X}_i \mid \mathbf{u}_i \sim \mathcal{N}(\boldsymbol{\mu} + \boldsymbol{\Lambda}\mathbf{u}_i, \psi\mathbf{I}_p)$, we have

$$
\begin{aligned}
\log f(\mathbf{x}_i \mid \mathbf{u}_i) &= -\frac{p}{2}\log 2\pi - \frac{1}{2}\log |\psi\mathbf{I}_p| \\
&\quad - \frac{1}{2}(\mathbf{x}_i - \boldsymbol{\mu} - \boldsymbol{\Lambda}\mathbf{u}_i)'(\psi\mathbf{I}_p)^{-1}(\mathbf{x}_i - \boldsymbol{\mu} - \boldsymbol{\Lambda}\mathbf{u}_i) \\
&= -\frac{p}{2}\log 2\pi - \frac{p}{2}\log \psi - \frac{1}{2\psi}\big[(\mathbf{x}_i - \boldsymbol{\mu})'(\mathbf{x}_i - \boldsymbol{\mu}) + (\mathbf{x}_i - \boldsymbol{\mu})'\boldsymbol{\Lambda}\mathbf{u}_i \\
&\quad + \mathbf{u}_i'\boldsymbol{\Lambda}'(\mathbf{x}_i - \boldsymbol{\mu}) - \mathbf{u}_i'\boldsymbol{\Lambda}'\boldsymbol{\Lambda}\mathbf{u}_i\big] \\
&= -\frac{p}{2}\log 2\pi - \frac{p}{2}\log \psi - \frac{1}{2\psi}\operatorname{tr}\{(\mathbf{x}_i - \boldsymbol{\mu})(\mathbf{x}_i - \boldsymbol{\mu})'\} \\
&\quad + \frac{1}{\psi}\operatorname{tr}\{(\mathbf{x}_i - \boldsymbol{\mu})'\boldsymbol{\Lambda}\mathbf{u}_i\} - \frac{1}{2\psi}\operatorname{tr}\{\boldsymbol{\Lambda}'\boldsymbol{\Lambda}\mathbf{u}_i\mathbf{u}_i'\}.
\end{aligned}
$$

Now, the complete-data log-likelihood can be written

$$
\begin{aligned}
l_\mathrm{c}(\boldsymbol{\mu}, \boldsymbol{\Lambda}, \boldsymbol{\Psi}) &= \sum_{i=1}^{n} \log\left[f(\mathbf{x}_i \mid \mathbf{u}_i)f(\mathbf{u}_i)\right] \\
&= C - \frac{np}{2}\log \psi - \frac{1}{2\psi}\operatorname{tr}\left\{\sum_{i=1}^{n}(\mathbf{x}_i - \boldsymbol{\mu})(\mathbf{x}_i - \boldsymbol{\mu})'\right\} \\
&\quad + \frac{1}{\psi}\sum_{i=1}^{n}\operatorname{tr}\{(\mathbf{x}_i - \boldsymbol{\mu})'\boldsymbol{\Lambda}\mathbf{u}_i\} - \frac{1}{2\psi}\operatorname{tr}\left\{\boldsymbol{\Lambda}'\boldsymbol{\Lambda}\sum_{i=1}^{n}\mathbf{u}_i\mathbf{u}_i'\right\},
\end{aligned}
$$

where C is constant with respect to $\boldsymbol{\mu}$, $\boldsymbol{\Lambda}$, and ψ.

Consider the joint distribution

$$
\begin{bmatrix} \mathbf{X}_i \\ \mathbf{U}_i \end{bmatrix} \sim \mathcal{N}\left(\begin{bmatrix} \boldsymbol{\mu} \\ \mathbf{0} \end{bmatrix}, \begin{bmatrix} \boldsymbol{\Lambda}\boldsymbol{\Lambda}' + \psi\mathbf{I}_p & \boldsymbol{\Lambda} \\ \boldsymbol{\Lambda}' & \mathbf{I}_q \end{bmatrix}\right).
$$

It follows that
$$\mathbb{E}[\mathbf{U}_i \mid \mathbf{x}_i] = \boldsymbol{\beta}(\mathbf{x}_i - \boldsymbol{\mu}), \tag{6.10}$$
where $\boldsymbol{\beta} = \boldsymbol{\Lambda}'(\boldsymbol{\Lambda}\boldsymbol{\Lambda}' + \psi\mathbf{I}_p)^{-1}$, and

$$\begin{aligned}
\mathbb{E}[\mathbf{U}_i\mathbf{U}_i' \mid \mathbf{x}_i] &= \mathbb{V}\mathrm{ar}[\mathbf{U}_i \mid \mathbf{x}_i] + \mathbb{E}[\mathbf{U}_i \mid \mathbf{x}_i]\mathbb{E}[\mathbf{U}_i \mid \mathbf{x}_i]' \\
&= \mathbf{I}_q - \boldsymbol{\beta}\boldsymbol{\Lambda} + \boldsymbol{\beta}(\mathbf{x}_i - \boldsymbol{\mu})(\mathbf{x}_i - \boldsymbol{\mu})'\boldsymbol{\beta}'.
\end{aligned} \tag{6.11}$$

Therefore, noting that $\hat{\boldsymbol{\mu}} = \bar{\mathbf{x}}$ and that we are conditioning on the current parameter estimates, the expected value of the complete-data log-likelihood can be written

$$\begin{aligned}
Q(\boldsymbol{\Lambda}, \psi) &= C - \frac{np}{2}\log\psi - \frac{1}{2\psi}\mathrm{tr}\left\{\sum_{i=1}^{n}(\mathbf{x}_i - \bar{\mathbf{x}})(\mathbf{x}_i - \bar{\mathbf{x}})'\right\} \\
&\quad + \frac{1}{\psi}\sum_{i=1}^{n}\mathrm{tr}\left\{(\mathbf{x}_i - \bar{\mathbf{x}})'\boldsymbol{\Lambda}\mathbb{E}\left[\mathbf{U}_i \mid \mathbf{x}_i\right]\right\} \\
&\quad - \frac{1}{2\psi}\mathrm{tr}\left\{\boldsymbol{\Lambda}'\boldsymbol{\Lambda}\sum_{i=1}^{n}\mathbb{E}[\mathbf{U}_i\mathbf{U}_i' \mid \mathbf{x}_{i,}]\right\} \\
&= C - \frac{np}{2}\log\psi - \frac{n}{2\psi}\mathrm{tr}\{\mathbf{S}_{\bar{\mathbf{x}}}\} + \frac{n}{\psi}\mathrm{tr}\{\boldsymbol{\Lambda}\hat{\boldsymbol{\beta}}\mathbf{S}_{\bar{\mathbf{x}}}\} - \frac{n}{2\psi}\mathrm{tr}\{\boldsymbol{\Lambda}'\boldsymbol{\Lambda}\boldsymbol{\Theta}\},
\end{aligned}$$

where $\boldsymbol{\Theta} = \mathbf{I}_q - \hat{\boldsymbol{\beta}}\hat{\boldsymbol{\Lambda}} + \hat{\boldsymbol{\beta}}\mathbf{S}_{\bar{\mathbf{x}}}\hat{\boldsymbol{\beta}}'$ is a symmetric $q \times q$ matrix, $\hat{\boldsymbol{\beta}} = \hat{\boldsymbol{\Lambda}}'(\hat{\boldsymbol{\Lambda}}\hat{\boldsymbol{\Lambda}}' + \hat{\psi}\mathbf{I}_p)^{-1}$, and

$$\mathbf{S}_{\bar{\mathbf{x}}} = \frac{1}{n}\sum_{i=1}^{n}(\mathbf{x}_i - \bar{\mathbf{x}})(\mathbf{x}_i - \bar{\mathbf{x}})' \tag{6.12}$$

can be thought of as the sample, or observed, covariance matrix.

6.4.3 M-step

Differentiating Q with respect to $\boldsymbol{\Lambda}$ and ψ, respectively, gives the score functions

$$\begin{aligned}
S_1(\boldsymbol{\Lambda}, \psi) = \frac{\partial Q}{\partial \boldsymbol{\Lambda}} &= \frac{n}{\psi}\frac{\partial}{\partial\boldsymbol{\Lambda}}\mathrm{tr}\{\boldsymbol{\Lambda}\hat{\boldsymbol{\beta}}\mathbf{S}_{\bar{\mathbf{x}}}\} - \frac{n}{2\psi}\frac{\partial}{\partial\boldsymbol{\Lambda}}\mathrm{tr}\{\boldsymbol{\Lambda}'\boldsymbol{\Lambda}\boldsymbol{\Theta}\} \\
&= \frac{n}{\psi}\mathbf{S}_{\bar{\mathbf{x}}}'\hat{\boldsymbol{\beta}}' - \frac{n}{2\psi}(2\boldsymbol{\Lambda}\boldsymbol{\Theta}) = \frac{n}{\psi}(\mathbf{S}_{\bar{\mathbf{x}}}\hat{\boldsymbol{\beta}}' - \boldsymbol{\Lambda}\boldsymbol{\Theta}),
\end{aligned}$$

and

$$S_2(\mathbf{\Lambda}, \psi) = \frac{\partial Q}{\partial \psi^{-1}}$$

$$= \frac{np\psi}{2} - \frac{n}{2}\operatorname{tr}\{\mathbf{S}_{\bar{\mathbf{x}}}\} + n\operatorname{tr}\{\mathbf{\Lambda}\hat{\beta}\mathbf{S}_{\bar{\mathbf{x}}}\} - \frac{n}{2}\operatorname{tr}\{\mathbf{\Lambda}'\mathbf{\Lambda}\mathbf{\Theta}\}$$

$$= \frac{n}{2}\left(p\psi - \operatorname{tr}\{\mathbf{S}_{\bar{\mathbf{x}}} - 2\mathbf{\Lambda}\hat{\beta}\mathbf{S}_{\bar{\mathbf{x}}} + \mathbf{\Lambda}'\mathbf{\Lambda}\mathbf{\Theta}\}\right).$$

Solving the equations $S_1(\hat{\mathbf{\Lambda}}^{\text{new}}, \hat{\psi}^{\text{new}}) = \mathbf{0}$ and $S_2(\hat{\mathbf{\Lambda}}^{\text{new}}, \hat{\psi}^{\text{new}}) = 0$ gives

$$\hat{\mathbf{\Lambda}}^{\text{new}} = \mathbf{S}_{\bar{\mathbf{x}}}\hat{\beta}'\mathbf{\Theta}^{-1},$$

$$\hat{\psi}^{\text{new}} = \frac{1}{p}\operatorname{tr}\{\mathbf{S}_{\bar{\mathbf{x}}} - 2\hat{\mathbf{\Lambda}}^{\text{new}}\hat{\beta}\mathbf{S}_{\bar{\mathbf{x}}} + (\hat{\mathbf{\Lambda}}^{\text{new}})'\hat{\mathbf{\Lambda}}^{\text{new}}\mathbf{\Theta}\}$$

$$= \frac{1}{p}\operatorname{tr}\{\mathbf{S}_{\bar{\mathbf{x}}} - 2\hat{\mathbf{\Lambda}}^{\text{new}}\hat{\beta}\mathbf{S}_{\bar{\mathbf{x}}} + (\mathbf{S}_{\bar{\mathbf{x}}}\hat{\beta}'\mathbf{\Theta}^{-1})'\hat{\mathbf{\Lambda}}^{\text{new}}\mathbf{\Theta}\}$$

$$= \frac{1}{p}\operatorname{tr}\{\mathbf{S}_{\bar{\mathbf{x}}} - \hat{\mathbf{\Lambda}}^{\text{new}}\hat{\beta}\mathbf{S}_{\bar{\mathbf{x}}}\}.$$

The matrix results used to compute these score functions, and used elsewhere in this section, are listed in Appendix C.

6.4.4 Woodbury Identity

On each iteration of the EM algorithm for PPCA, the $p \times p$ matrix $(\hat{\mathbf{\Lambda}}\hat{\mathbf{\Lambda}}' + \hat{\psi}\mathbf{I}_p)^{-1}$ needs to be computed. Computing this matrix inverse can be computationally expensive, especially for larger values of p. The Woodbury identity (Woodbury, 1950) can be used in such situations to avoid inversion of non-diagonal $p \times p$ matrices. For an $m \times m$ matrix $\mathbf{A}$, an $m \times k$ matrix $\mathbf{U}$, a $k \times k$ matrix $\mathbf{C}$, and a $k \times m$ matrix $\mathbf{V}$, the Woodbury identity is

$$(\mathbf{A} + \mathbf{U}\mathbf{C}\mathbf{V})^{-1} = \mathbf{A}^{-1} - \mathbf{A}^{-1}\mathbf{U}(\mathbf{C}^{-1} + \mathbf{V}\mathbf{A}^{-1}\mathbf{U})^{-1}\mathbf{V}\mathbf{A}^{-1}.$$

Setting $\mathbf{U} = \hat{\mathbf{\Lambda}}$, $\mathbf{V} = \hat{\mathbf{\Lambda}}'$, $\mathbf{A} = \hat{\psi}\mathbf{I}_p$, and $\mathbf{C} = \mathbf{I}_q$ gives

$$\begin{aligned}(\hat{\psi}\mathbf{I}_p + \hat{\mathbf{\Lambda}}\hat{\mathbf{\Lambda}}')^{-1} &= \frac{1}{\hat{\psi}}\mathbf{I}_p - \frac{1}{\hat{\psi}}\mathbf{\Lambda}\left(\mathbf{I}_q + \frac{1}{\hat{\psi}}\mathbf{\Lambda}'\mathbf{\Lambda}\right)^{-1}\mathbf{\Lambda}'\frac{1}{\hat{\psi}} \\ &= \frac{1}{\hat{\psi}}\mathbf{I}_p - \frac{1}{\hat{\psi}}\mathbf{\Lambda}\left(\hat{\psi}\mathbf{I}_q + \mathbf{\Lambda}'\mathbf{\Lambda}\right)^{-1}\mathbf{\Lambda}',\end{aligned} \qquad (6.13)$$

which can be used to speed up the EM algorithm for the PPCA model. The left-hand side of (6.13) requires inversion of a $p \times p$

matrix but the right-hand side leaves only a $q \times q$ matrix and some diagonal matrices to be inverted. A related identity for the determinant of the covariance matrix,

$$
\begin{aligned}
|\hat{\boldsymbol{\Lambda}}\hat{\boldsymbol{\Lambda}}' + \hat{\psi}\mathbf{I}_p| &= \frac{|\hat{\psi}\mathbf{I}_p|}{|\mathbf{I}_q - \boldsymbol{\Lambda}'(\boldsymbol{\Lambda}\boldsymbol{\Lambda}' + \hat{\psi}\mathbf{I}_p)^{-1}\boldsymbol{\Lambda}|} \\
&= \frac{\hat{\psi}^p}{|\mathbf{I}_q - \boldsymbol{\Lambda}'(\boldsymbol{\Lambda}\boldsymbol{\Lambda}' + \hat{\psi}\mathbf{I}_p)^{-1}\boldsymbol{\Lambda}|},
\end{aligned}
\tag{6.14}
$$

is also helpful in computation of the component densities. Identities (6.13) and (6.14) give an especially significant computational advantage when p is large and $q \ll p$.

6.4.5 Initialization

Similar to McNicholas and Murphy (2008), the parameters $\hat{\boldsymbol{\Lambda}}$ and $\hat{\psi}$ can be initialized in several ways, including via eigen-decomposition of $\mathbf{S}_{\bar{\mathbf{x}}}$. Specifically, $\mathbf{S}_{\bar{\mathbf{x}}}$ is computed and then it is eigen-decomposed to give

$$
\mathbf{S}_{\bar{\mathbf{x}}} = \mathbf{P}\mathbf{D}\mathbf{P}^{-1},
$$

and $\hat{\boldsymbol{\Lambda}}$ is initialized using

$$
\hat{\boldsymbol{\Lambda}} = \mathbf{d}\mathbf{P},
$$

where $\mathbf{d}$ is the element-wise square root of the diagonal of $\mathbf{D}$. The initial value of ψ can be taken to be

$$
\hat{\psi} = \frac{1}{p}\operatorname{tr}\{\mathbf{S}_{\bar{\mathbf{x}}} - (\mathbf{d}\mathbf{P})(\mathbf{d}\mathbf{P})'\}.
$$

6.4.6 Stopping Rule

There are several options for stopping an EM algorithm. One popular approach is to stop an EM algorithm based on lack of progress in the log-likelihood, i.e., stopping the algorithm when

$$
l^{(k+1)} - l^{(k)} < \epsilon,
\tag{6.15}
$$

for ϵ small, where $l^{(k)}$ is the (observed) log-likelihood value from iteration k. This stopping rule can work very well when the log-likelihood increases and then plateaus at the maximum likelihood estimate. However, likelihood values do not necessarily behave this way and so it can be worth considering alternatives — see Chapter 2 of McNicholas (2016a) for an illustrated discussion.

As McNicholas (2016a) points out, Böhning et al. (1994), Lindsay (1995), and McNicholas et al. (2010) consider convergence criteria based on Aitken's acceleration Aitken (1926). Aitken's acceleration at iteration k is

$$a^{(k)} = \frac{l^{(k+1)} - l^{(k)}}{l^{(k)} - l^{(k-1)}}, \tag{6.16}$$

and an asymptotic estimate of the log-likelihood at iteration $k + 1$ can be computed via

$$l_\infty^{(k+1)} = l^{(k)} + \frac{l^{(k+1)} - l^{(k)}}{1 - a^{(k)}}. \tag{6.17}$$

Note that "asymptotic" here refers to the iteration number so that (6.17) can be interpreted as an estimate, at iteration $k + 1$, of the ultimate value of the log-likelihood. Following McNicholas et al. (2010), the algorithm can be considered to have converged when

$$0 < l_\infty^{(k+1)} - l^{(k)} < \epsilon. \tag{6.18}$$

6.4.7 Implementing the EM Algorithm for PPCA

The EM algorithm for PPCA is implemented in two steps. First, we define some helper functions in the following code block. They are used by the EM algorithm function. Specifically, we have implemented a function for each of the identities (6.13) and (6.14) along with functions for each of the equations (6.8) and (6.16). Doing this helps keep the code modular, which facilitates testing some of the more complex mathematics in the algorithm.

```
## EM Algorithm for PPCA -- Helper functions

function wbiinv(q, p, psi, L)
    ## Woodbury Indentity eq 6.13
    Lt = transpose(L)
    Iq = Matrix{Float64}(I, q, q)
    Ip = Matrix{Float64}(I, p, p)
    psi2 = /(1, psi)
    m1 = *(psi2, Ip)
    m2c = +( *(psi, Iq), *(Lt, L) )
    m2 = *(psi2, L, inv(m2c), Lt)
    return( -(m1, m2) )
end
```

```
function wbidet(q, p, psi, L)
  ## Woodbury Indentity eq 6.14
  Iq = Matrix{Float64}(I, q, q)
  num = psi^p
  den1 = *(transpose(L), wbiinv(q, p, psi, L), L)
  den = det(-(Iq, den1))
  return( /(num, den) )
end

function ppca_ll(n, q, p, psi, L, S)
    ## log likelihood eq 6.8
    n2 = /(n, 2)
    c = /( *(n, p, log(*(2, pi))), 2)
    l1 = *(n2, log(wbidet(q, p, psi, L)))
    l2 = *(n2, tr(*(S, wbiinv(q, p, psi, L))))
    return( *(-1, +(c, l1, l2)) )
end

function cov_mat(X)
    n,p = size(X)
    mux = mean(X, dims = 1)
    X_c = .-(X, mux)
    res = *( /(1,n), *(transpose(X_c), X_c))
    return(res)
end

function aitkenaccel(l1, l2, l3, tol)
    ## l1 = l(k+1), l2 = l(k), l3 = l(k-1)
    conv = false
    num = -(l1, l2)
    den = -(l2, l3)
    ## eq 6.16
    ak = /(num, den)
    if ak <= 1.0
        c1 = -(l1, l2)
        c2 = -(1.0, ak)
        ## eq 6.17
        l_inf = +(l2, /(c2, c1))
        c3 = -(l_inf, l2)
        if  0.0 < c3 < tol
            conv = true
        end
    end
    return conv
end

function ppca_fparams(p, q)
    ## number of free parameters in the ppca model
    c1 = *(p, q)
    c2 = *(-1, q, -(q, 1), 0.5)
    return( +(c1, c2, 1) )
end

function BiC(ll, n, q, ρ)
    ## name does not conflict with StatsBase
    ## eq 6.19
    c1 = *(2, ll)
    c2 = *(ρ, log(n))
    return( -(c1, c2) )
end
```

The `ppca()` function, illustrated in the following code block, carries out the EM algorithm described previously. It uses the eigen-decomposition of the sample covariance matrix to initialize Λ, which is used in turn to initialize ψ. A `while` loop is used to iteratively perform the (post-) E-step updates for β and Θ as well as the M-step updates of Λ and ψ. After the M-step, the log-likelihood is calculated and its value is stored in an array. The values in the array are used to calculate the Aitken acceleration to check convergence. The Woodbury identity is used to speed up the beta updates and the log-likelihood computations. The function returns a Julia dictionary with the PPCA results. The results include the number of iterations the EM algorithm ran, the array of log-likelihoods, the model parameters, the number of latent components q, the standard deviations from the eigen-decomposition, the orthonormal coefficients, and the projections of the data onto the latent space.

```julia
using LinearAlgebra, Random
Random.seed!(429)
include("chp6_ppca_functions.jl")

## EM Function for PPCA
function ppca(X; q = 2, thresh = 1e-5, maxit = 1e5)

    Iq = Matrix{Float64}(I, q, q)
    qI = *(q, Iq)
    n,p = size(X)
    ## eigfact has eval/evec smallest to largest
    ind = p:-1:(p-q+1)
    Sx = cov_mat(X)
    D = eigen(Sx)
    d = diagm(0 => map(sqrt, D.values[ind]))
    P = D.vectors[:, ind]

    ## initialize parameters
    L = *(P, d) ## pxq
    psi = *( /(1, p), tr( -( Sx, *(L, transpose(L)) ) ) )
    B = zeros(q,p)
    T = zeros(q,q)
    conv = true
    iter = 1
    ll = zeros(100)
    ll[1] = 1e4

    ## while not converged
    while(conv)
        ## above eq 6.12
        B = *(transpose(L), wbiinv(q, p, psi, L))  ## qxp
        T = -(qI, +( *(B,L), *(B, Sx, transpose(B)) ) ) ## qxq
        ## sec 6.4.3 - update Lambda_new
        L_new = *(Sx, transpose(B), inv(T)) ## pxq
        ## update psi_new
        psi_new = *( /(1, p), tr( -(Sx, *(L_new, B, Sx) ) )) #num
```

```
    iter += 1

    if iter > maxit
        conv = false
        println("ppca() while loop went over maxit parameter. iter =
          $iter")
    else
      ## stores at most the 100 most recent ll values
      if iter <= 100
        ll[iter]=ppca_ll(n, q, p, psi_new, L_new, Sx)
      else
        ll = circshift(ll, -1)
        ll[100] = ppca_ll(n, q, p, psi_new, L_new, Sx)
      end
      if 2 < iter < 101
          ## scales the threshold by the latest ll value
          thresh2 = *(-1, ll[iter], thresh)
          if aitkenaccel(ll[iter], ll[(iter-1)], ll[(iter-2)], thresh2)
           conv = false
          end
      else
        thresh2 = *(-1, ll[100], thresh)
        if aitkenaccel(ll[100], ll[(99)], ll[(98)], thresh2)
          conv = false
        end
      end
    end ## if maxit
    L = L_new
    psi = psi_new
  end ## while

  ## orthonormal coefficients
  coef = svd(L).U
  ## projections
  proj = *(X, coef)

  if iter <= 100
    resize!(ll, iter)
  end
  fp = ppca_fparams(p, q)

  bic_res = BiC(ll[end], n, q, fp)

  return(Dict(
      "iter" => iter,
      "ll" => ll,
      "beta" => B,
      "theta" => T,
      "lambda" => L,
      "psi" => psi,
      "q" => q,
      "sd" => diag(d),
      "coef" => coef,
      "proj" => proj,
      "bic" => bic_res
    ))
end

ppca1 = ppca(crab_mat_c, q=3, maxit = 1e3)
```

It should be noted that the **Λ** matrix is neither orthonormal nor composed of eigenvectors (Tipping and Bishop, 1999b). To get orthonormal principal components from it, a singular value decomposition is performed on **Λ** and the left singular vectors are used as principal component loadings. The projections are plotted in Figure 6.3. With three latent components, we see a clear separation between the crab species.

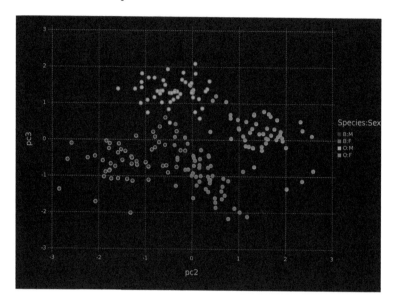

Figure 6.3 Scallerplot depicting the second and third principal components from the PPCA model with three latent components for the `crabs` data, coloured by sex and species.

6.4.8 Comments

The choice of the number of latent components is an important consideration in PPCA. One approach is to choose the number of latent components that captures a certain proportion of the variation in the data. Lopes and West (2004) carry out simulation studies showing that the Bayesian information criterion (BIC; Schwarz, 1978) can be effective for selection of the number of latent factors in a factor analysis model, and a similar approach can be followed for PPCA. The BIC is given by

$$\text{BIC} = 2l(\hat{\boldsymbol{\vartheta}}) - \rho \log n, \tag{6.19}$$

where $\hat{\boldsymbol{\vartheta}}$ is the maximum likelihood estimate of $\boldsymbol{\vartheta}$, $l(\hat{\boldsymbol{\vartheta}})$ is the maximized (observed) log-likelihood, and ρ is the number of free parameters in the model. The BIC values for PPCA models with different numbers of latent components run on the `crabs` data are displayed in figure 6.4. The clear choice is the model with three latent components, which has the largest BIC value based on equation 6.19. Clearly this model is capturing the important information in the third principal component, visualized in figure 6.3.

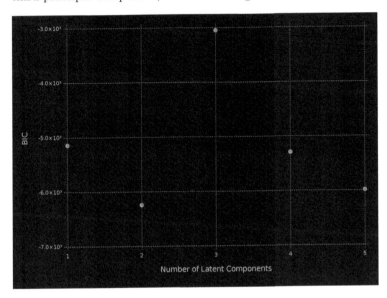

Figure 6.4   BIC values for the PPCA models with different numbers of latent components for the `crabs` data.

The goal of PPCA is not to necessarily give better results than PCA, but to permit a range of extensions. The probabilistic formulation of the PPCA model allows these extensions, one of which will be illustrated in Section 6.6. Because the PPCA model has a likelihood formulation, it can easily be compared to other probabilistic models, it may be formulated to work with missing data, and adapted for use as a Bayesian method. Noted that, as $\psi \to 0$, the PPCA solution converges to the PCA solution (see Tipping and Bishop, 1999b, for details).

6.5 K-MEANS CLUSTERING

The k-means clustering technique iteratively finds k cluster centres and assigns each observation to the nearest cluster centre. For k-means clustering, these cluster centres are simply the means. The result of k-means clustering is effectively to fit k circles of equal radius to data, where each circle centre corresponds to one of the k means. Of course, for $p > 2$ dimensions, we have p-spheres rather than circles. Consider the k-means clustering of the x2 data in the following code block. Recall that the x2 data is just a mixture of three bivariate Gaussian distributions (Figure 6.5).

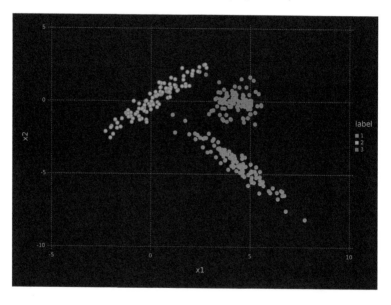

Figure 6.5 Scallerplot depicting the x2 data, coloured by class.

The x2 data present an interesting example of simulated data where classes from the generative model, i.e., a three-component Gaussian mixture model, cannot quite be taken as the correct result. Specifically, an effective clustering procedure would surely put one of the green-coloured points in Figure 6.5 in the same cluster as the yellow points.

In the following code block, the choice of k is made using the elbow method, where one plots k against the total cost, i.e., the total within-cluster sum of squares, and chooses the value of k corresponding to the "elbow" in the plot. Note that the total within-cluster sum of squares is

$$\sum_{g=1}^{k} \sum_{\mathbf{x}_i \in C_g} (\mathbf{x}_i - \bar{\mathbf{x}}_g)^2,$$

where C_g is the gth cluster and $\bar{\mathbf{x}}_g$ is its mean.

```
using DataFrames, Clustering, Gadfly, Random
Random.seed!(429)

mean_x2 = mean(x2_mat, dims=1)
## mean center the cols
x2_mat_c = x2_mat .- mean_x2
N = size(x2_mat_c)[1]

## kmeans() - each column of X is a sample - requires reshaping x2
x2_mat_t = reshape(x2_mat_c, (2,N))

## Create data for elbow plot
k = 2:8
df_elbow = DataFrame(k = Vector{Int64}(), tot_cost = Vector{Float64}())
for i in k
    tmp = [i, kmeans(x2_mat_t, i; maxiter=10, init=:kmpp).totalcost ]
    push!(df_elbow, tmp)
end

## create elbow plot
p_km_elbow = plot(df_elbow, x = :k, y = :tot_cost, Geom.point, Geom.line,
    Guide.xlabel("k"), Guide.ylabel("Total Within Cluster SS"),
    Coord.Cartesian(xmin = 1.95), Guide.xticks(ticks = collect(2:8)))
```

From Figure 6.6, it is clear that the elbow occurs at $k = 3$. The performance of k-means clustering on this data is not particularly good (Figure 6.7), which is not surprising. The reason this result is not surprising is that, as mentioned before, k-means clustering will essentially fit k circles — for the x2 data, two of the three clusters are long ellipses.

It is of interest to consider the solution for $k = 2$ (Figure 6.8) because it further illustrates the reliance of k-means clustering on clusters that are approximately circles. An example where k-means should, and does, work well is shown in Figure 6.9, where the clusters are well described by fitted circles of equal radius. It is also worth noting that k-means clustering works well for trivial clustering problems where there is significant spatial separation between each cluster, regardless of the cluster shapes.

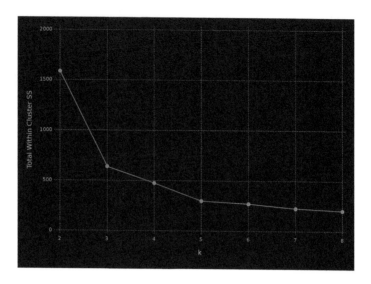

Figure 6.6 Elbow plot for selecting k in the k-means clustering of the x2 data.

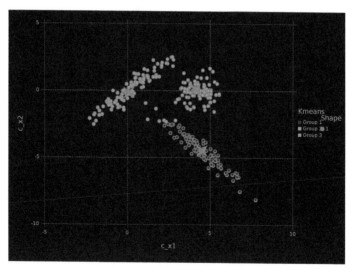

Figure 6.7 Scallerplot depicting the x2 data, coloured by the k-means clustering solution for $k = 3$, with red stars marking the cluster means.

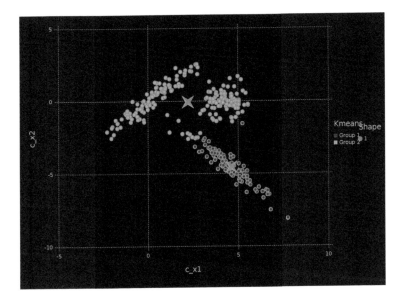

Figure 6.8 Scallerplot depicting the x2 data, coloured by the k-means clustering solution for $k = 2$, with red stars marking the cluster means.

6.6 MIXTURE OF PROBABILISTIC PRINCIPAL COMPONENTS ANALYZERS

6.6.1 Model

Building on the PPCA model, Tipping and Bishop (1999a) introduced the the mixture of probabilistic principal components analyzers (MPPCA) model. It can model complex structures in a dataset using a combination of local PPCA models. As PPCA provides a reduced dimensional representation of the data, MPPCA models work well in high-dimensional clustering, density estimation and classification problems.

Analogous to the PPCA model (Section 6.3), the MPPCA model assumes that

$$\mathbf{X}_i = \boldsymbol{\mu}_g + \boldsymbol{\Lambda}_g \mathbf{U}_{ig} + \boldsymbol{\epsilon}_{ig} \qquad (6.20)$$

with probability π_g, for $i = 1, \ldots, n$ and $g = 1, \ldots, G$, where $\boldsymbol{\Lambda}_g$ is a $p \times q$ matrix of loadings, the $\mathbf{U}_{ig}$ are independently $N(\mathbf{0}, \mathbf{I}_q)$ and are independent of the $\boldsymbol{\epsilon}_{ig}$, which are independently $N(\mathbf{0}, \psi_g \mathbf{I}_p)$, where

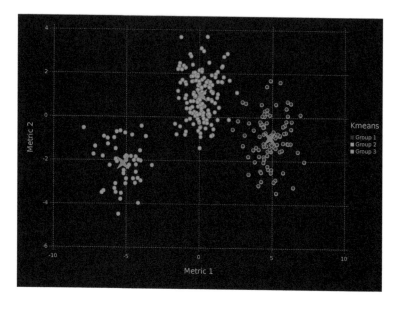

Figure 6.9 Scallerplot depicting a dataset where k-means works well, coloured by the k-means clustering solution for $k = 3$, with red stars marking the cluster means.

$\psi_g \in \mathbb{R}^+$. It follows that the density of $\mathbf{X}_i$ from the MPPCA model is

$$f(\mathbf{x}_i \mid \boldsymbol{\vartheta}) = \sum_{g=1}^{G} \pi_g \phi(\mathbf{x}_i \mid \boldsymbol{\mu}_g, \boldsymbol{\Lambda}_g \boldsymbol{\Lambda}_g' + \psi_g \mathbf{I}_p), \qquad (6.21)$$

where $\boldsymbol{\vartheta}$ denotes the model parameters.

6.6.2 Parameter Estimation

Overview

Parameter estimation for the MPPCA model can be carried out using an alternating expectation-conditional maximization (AECM) algorithm (Meng and van Dyk, 1997). The expectation-conditional maximization (ECM) algorithm (Meng and Rubin, 1993) is a variant of the EM algorithm that replaces the M-step by a series of conditional maximization steps. The AECM algorithm allows a different specification of complete-data for each conditional maximization step. Accordingly, the AECM algorithm is suitable for the MPPCA model, where there are two sources of missing data: the unknown component membership labels and the latent components (or factors) $\mathbf{u}_{ig}$, for $i = 1, \ldots, n$ and $g = 1, \ldots, G$. Details of fitting the AECM algorithm for the more general mixture of factor analyzers model are given by McLachlan and Peel (2000), and parameter estimation for several related models is discussed by McNicholas and Murphy (2008, 2010) and McNicholas et al. (2010).

AECM Algorithm: First Stage

As usual, denote by $\mathbf{z}_1, \ldots, \mathbf{z}_n$ the unobserved component membership labels, where $z_{ig} = 1$ if observation i belongs to component g and $z_{ig} = 0$ otherwise. At the first stage of the AECM algorithm, the complete-data are taken to be the observed $\mathbf{x}_1, \ldots, \mathbf{x}_n$ together with the unobserved $\mathbf{z}_1, \ldots, \mathbf{z}_n$, and the parameters π_g and $\boldsymbol{\mu}_g$ are estimated, for $g = 1, \ldots, G$. The complete-data log-likelihood is

$$l_1 = \sum_{i=1}^{n} \sum_{g=1}^{G} z_{ig} \log \left[\pi_g \phi(\mathbf{x}_i \mid \boldsymbol{\mu}_g, \boldsymbol{\Lambda}_g \boldsymbol{\Lambda}'_g + \psi_g \mathbf{I}_p) \right], \qquad (6.22)$$

and the (conditional) expected values of the component membership labels are given by

$$\hat{z}_{ig} = \frac{\hat{\pi}_g \phi(\mathbf{x}_i \mid \hat{\boldsymbol{\mu}}_g, \hat{\boldsymbol{\Lambda}}_g \hat{\boldsymbol{\Lambda}}'_g + \hat{\psi}_g \mathbf{I}_p)}{\sum_{h=1}^{G} \hat{\pi}_h \phi(\mathbf{x}_i \mid \hat{\boldsymbol{\mu}}_h, \hat{\boldsymbol{\Lambda}}_h \hat{\boldsymbol{\Lambda}}'_h + \hat{\psi}_h \mathbf{I}_p)}, \qquad (6.23)$$

for $i = 1, \ldots, n$ and $g = 1, \ldots, G$.

Using the expected values given by (6.23) within (6.22), the expected value of the complete-data log-likelihood at the first stage is

$$\begin{aligned} Q_1 &= \sum_{i=1}^{n} \sum_{g=1}^{G} \hat{z}_{ig} \left[\log \pi_g + \log \phi(\mathbf{x}_i \mid \boldsymbol{\mu}_g, \boldsymbol{\Lambda}_g \boldsymbol{\Lambda}'_g + \psi_g \mathbf{I}_p) \right] \\ &= \sum_{g=1}^{G} n_g \log \pi_g - \frac{np}{2} \log 2\pi - \sum_{g=1}^{G} \frac{n_g}{2} \log |\boldsymbol{\Lambda}_g \boldsymbol{\Lambda}'_g + \psi_g \mathbf{I}_p| \\ &\qquad\qquad - \sum_{g=1}^{G} \frac{n_g}{2} \operatorname{tr} \left\{ \mathbf{S}_g (\boldsymbol{\Lambda}_g \boldsymbol{\Lambda}'_g + \psi_g \mathbf{I}_p)^{-1} \right\}, \end{aligned}$$

where $n_g = \sum_{i=1}^{n} \hat{z}_{ig}$ and

$$S_g = \frac{1}{n_g} \sum_{i=1}^{n} \hat{z}_{ig} (\mathbf{x}_i - \boldsymbol{\mu}_g)(\mathbf{x}_i - \boldsymbol{\mu}_g)'. \tag{6.24}$$

Maximising Q_1 with respect to π_g and $\boldsymbol{\mu}_g$ yields

$$\hat{\pi}_g = \frac{n_g}{n} \qquad \text{and} \qquad \hat{\boldsymbol{\mu}}_g = \frac{\sum_{i=1}^{n} \hat{z}_{ig} \mathbf{x}_i}{\sum_{i=1}^{n} \hat{z}_{ig}}, \tag{6.25}$$

respectively.

AECM Algorithm: Second Stage

At the second stage of the AECM algorithm, the complete-data are taken to be the observed $\mathbf{x}_1, \ldots, \mathbf{x}_n$ together with the unobserved component membership labels $\mathbf{z}_1, \ldots, \mathbf{z}_n$ and the latent factors $\mathbf{u}_{ig}$, for $i = 1, \ldots, n$ and $g = 1, \ldots, G$, and the parameters $\boldsymbol{\Lambda}_g$ and ψ_g are estimated, for $g = 1, \ldots, G$. Proceeding in an analogous fashion to the EM algorithm for the factor analysis model (Section 6.4), the complete-data log-likelihood is given by

$$l_2 = \sum_{i=1}^{n} \sum_{g=1}^{G} \hat{z}_{ig} \left[\log \pi_g + \log f(\mathbf{x}_i | \mathbf{u}_i) + \log f(\mathbf{u}_i) \right]$$

$$= C + \sum_{g=1}^{G} \left[-\frac{n_g p}{2} \log \psi_g - \frac{n_g}{2\psi_g} \operatorname{tr}\{S_g\} \right.$$

$$\left. + \frac{1}{\psi_g} \sum_{i=1}^{n} z_{ig} (\mathbf{x}_i - \boldsymbol{\mu}_g)' \boldsymbol{\Lambda}_g \mathbf{u}_i - \frac{1}{2\psi_g} \operatorname{tr}\left\{ \boldsymbol{\Lambda}_g' \boldsymbol{\Lambda}_g \sum_{i=1}^{n} z_{ig} \mathbf{u}_i \mathbf{u}_i' \right\} \right],$$

where C is constant with respect to $\boldsymbol{\Lambda}_g$ and $\boldsymbol{\Psi}_g$. Bearing in mind that we are conditioning on the current parameter estimates, and using expected values analogous to those in (6.10) and (6.11), the expected value of the complete-data log-likelihood can be written

$$Q_2 = C + \sum_{g=1}^{G} \left[-\frac{n_g p}{2} \log \psi_g - \frac{n_g}{2\psi_g} \operatorname{tr}\{\mathbf{S}_g\} \right.$$

$$+ \frac{1}{\psi_g} \sum_{i=1}^{n} \hat{z}_{ig}(\mathbf{x}_i - \hat{\boldsymbol{\mu}}_g)' \boldsymbol{\Lambda}_g \mathbb{E}[\mathbf{U}_{ig} \mid \mathbf{x}_i, z_{ig} = 1]$$

$$\left. - \frac{1}{2\psi_g} \operatorname{tr}\left\{ \boldsymbol{\Lambda}_g' \boldsymbol{\Lambda}_g \sum_{i=1}^{n} \hat{z}_{ig} \mathbb{E}[\mathbf{U}_{ig}\mathbf{U}_{ig}' \mid \mathbf{x}_i, z_{ig} = 1] \right\} \right]$$

$$= C + \frac{1}{2} \sum_{g=1}^{G} n_g \left[-p \log \psi_g - \frac{1}{\psi_g} \operatorname{tr}\{\mathbf{S}_g\} + \frac{2}{\psi_g} \operatorname{tr}\{\boldsymbol{\Lambda}_g \hat{\boldsymbol{\beta}}_g \mathbf{S}_g\} \right.$$

$$\left. - \frac{1}{\psi_g} \operatorname{tr}\{\boldsymbol{\Lambda}_g' \boldsymbol{\Lambda}_g \boldsymbol{\Theta}_g\} \right],$$

where

$$\hat{\boldsymbol{\beta}}_g = \hat{\boldsymbol{\Lambda}}_g' (\hat{\boldsymbol{\Lambda}}_g \hat{\boldsymbol{\Lambda}}_g' + \hat{\psi}_g \mathbf{I}_p)^{-1}$$

is a $q \times p$ matrix and

$$\boldsymbol{\Theta}_g = \mathbf{I}_q - \hat{\boldsymbol{\beta}}_g \hat{\boldsymbol{\Lambda}}_g + \hat{\boldsymbol{\beta}}_g \mathbf{S}_g \hat{\boldsymbol{\beta}}_g'$$

is a symmetric $q \times q$ matrix. Note that $\hat{\boldsymbol{\mu}}_g$ replaces $\boldsymbol{\mu}_g$ in $\mathbf{S}_g$, see (6.24).

Differentiating $Q_2(\boldsymbol{\Lambda}, \boldsymbol{\Psi})$ with respect to $\boldsymbol{\Lambda}_g$ and ψ_g^{-1}, respectively, gives the score functions

$$S_1(\boldsymbol{\Lambda}_g, \psi_g) = \frac{\partial Q_2}{\partial \boldsymbol{\Lambda}_g} = \frac{n_g}{\psi_g} \frac{\partial}{\partial \boldsymbol{\Lambda}_g} \operatorname{tr}\{\boldsymbol{\Lambda}_g \hat{\boldsymbol{\beta}}_g \mathbf{S}_g\} - \frac{n_g}{2\psi_g} \frac{\partial}{\partial \boldsymbol{\Lambda}_g} \operatorname{tr}\{\boldsymbol{\Lambda}_g' \boldsymbol{\Lambda}_g \boldsymbol{\Theta}_g\}$$

$$= \frac{n_g}{\psi_g} \mathbf{S}_g' \hat{\boldsymbol{\beta}}_g' - \frac{n_g}{2\psi_g} (2\boldsymbol{\Lambda}_g \boldsymbol{\Theta}_g)$$

$$= \frac{n_g}{\psi_g} (\mathbf{S}_g \hat{\boldsymbol{\beta}}_g' - \boldsymbol{\Lambda}_g \boldsymbol{\Theta}_g),$$

and

$$S_2(\boldsymbol{\Lambda}_g, \psi_g) = \frac{\partial Q}{\partial \psi_g^{-1}}$$

$$= \frac{n_g p \psi_g}{2} - \frac{n_g}{2} \operatorname{tr}\{\mathbf{S}_g\} + n_g \operatorname{tr}\{\boldsymbol{\Lambda}_g \hat{\boldsymbol{\beta}}_g \mathbf{S}_g\} - \frac{n_g}{2} \operatorname{tr}\{\boldsymbol{\Lambda}_g' \boldsymbol{\Lambda}_g \boldsymbol{\Theta}_g\}$$

$$= \frac{n_g}{2} \left(p\psi_g - \operatorname{tr}\{\mathbf{S}_g - 2\boldsymbol{\Lambda}_g \hat{\boldsymbol{\beta}}_g \mathbf{S}_g + \boldsymbol{\Lambda}_g' \boldsymbol{\Lambda}_g \boldsymbol{\Theta}_g\} \right).$$

Solving the equations $S_1(\hat{\mathbf{\Lambda}}_g^{\text{new}}, \hat{\psi}_g^{\text{new}}) = \mathbf{0}$ and $S_2(\hat{\mathbf{\Lambda}}_g^{\text{new}}, \hat{\psi}_g^{\text{new}}) = 0$ gives

$$\hat{\mathbf{\Lambda}}_g^{\text{new}} = \mathbf{S}_g \hat{\boldsymbol{\beta}}_g' \boldsymbol{\Theta}_g^{-1},$$

$$\hat{\psi}_g^{\text{new}} = \frac{1}{p} \text{tr}\{\mathbf{S}_g - 2\hat{\mathbf{\Lambda}}_g^{\text{new}} \hat{\boldsymbol{\beta}}_g \mathbf{S}_g + (\hat{\mathbf{\Lambda}}_g^{\text{new}})' \hat{\mathbf{\Lambda}}_g^{\text{new}} \boldsymbol{\Theta}_g\}$$

$$= \frac{1}{p} \text{tr}\{\mathbf{S}_g - 2\hat{\mathbf{\Lambda}}_g^{\text{new}} \hat{\boldsymbol{\beta}}_g \mathbf{S}_g + (\mathbf{S}_g \hat{\boldsymbol{\beta}}_g' \boldsymbol{\Theta}_g^{-1})' \hat{\mathbf{\Lambda}}_g^{\text{new}} \boldsymbol{\Theta}_g\}$$

$$= \frac{1}{p} \text{tr}\{\mathbf{S}_g - \hat{\mathbf{\Lambda}}_g^{\text{new}} \hat{\boldsymbol{\beta}}_g \mathbf{S}_g\}.$$

The matrix results used to compute these score functions, and used elsewhere in this section, are listed in Appendix C.

AECM Algorithm for the MPPCA Model

An AECM algorithm for the MPPCA model can now be presented.

AECM Algorithm for MPPCA

initialize $\hat{\mathbf{z}}_{ig}$
initialize $\hat{\pi}_g, \hat{\boldsymbol{\mu}}_g, \mathbf{S}_g, \hat{\mathbf{\Lambda}}_g, \hat{\psi}_g$
while convergence criterion not met
    update $\hat{\pi}_g, \hat{\boldsymbol{\mu}}_g$
    if not iteration 1
        update $\hat{\mathbf{z}}_{ig}$
    end
    compute $\mathbf{S}_g, \hat{\boldsymbol{\beta}}_g, \boldsymbol{\Theta}_g$
    update $\hat{\mathbf{\Lambda}}_g^{\text{new}}, \hat{\psi}_g^{\text{new}}$
    update $\hat{\mathbf{z}}_{ig}$
    check convergence criterion
    $\hat{\mathbf{\Lambda}}_g \leftarrow \hat{\mathbf{\Lambda}}_g^{\text{new}}, \hat{\psi}_g \leftarrow \hat{\psi}_g^{\text{new}}$
end

In the following code block, we define some helper functions for the MPPCA model.

```
## MPPCA helper functions
using LinearAlgebra

function sigmaG(mu, xmat, Z)
    res = Dict{Int, Array}()
    N,g = size(Z)
```

```
    c1 = ./(1, sum(Z, dims = 1))
    for i = 1:g
        xmu = .-(xmat, transpose(mu[:,i]))
        zxmu = .*(Z[:,i], xmu)
        res_g = *(c1[i], *(transpose(zxmu), zxmu))
        push!(res, i=>res_g)
    end
    return res
end

function muG(g, xmat, Z)
    N, p = size(xmat)
    mu = zeros(p, g)
    for i in 1:g
        num = sum(.*(Z[:,i], xmat), dims = 1)
        den = sum(Z[:, i])
        mu[:, i] = /(num, den)
    end
    return mu
end

## initializations
function init_LambdaG(q, g, Sx)
    res = Dict{Int, Array}()
    for i = 1:g
        p = size(Sx[i], 1)
        ind = p:-1:(p-q+1)
        D = eigen(Sx[i])
        d = diagm(0 => map(sqrt, D.values[ind]))
        P = D.vectors[:, ind]
        L = *(P,d)
        push!(res, i => L)
    end
    return res
end

function init_psi(g, Sx, L)
    res = Dict{Int, Float64}()
    for i = 1:g
        p = size(Sx[i], 1)
        psi = *(/(1, p), tr(-(Sx[i], *(L[i], transpose(L[i])))))
        push!(res, i=>psi)
    end
    return res
end

function init_dict0(g, r, c)
    res = Dict{Int, Array}()
    for i = 1:g
        push!(res, i=> zeros(r, c))
    end
    return res
end

## Updates
function update_B(q, p, psig, Lg)
    res = Dict{Int, Array}()
    g = length(psig)
    for i=1:g
        B = *(transpose(Lg[i]), wbiinv(q, p, psig[i], Lg[i]))
        push!(res, i=>B)
    end
```

```julia
        return res
end

function update_T(q, Bg, Lg, Sg)
    res = Dict{Int, Array}()
    Iq = Matrix{Float64}(I, q, q)
    qI = *(q, Iq)
    g = length(Bg)
    for i =1:g
        T = -(qI, +(*(Bg[i], Lg[i]), *(Bg[i], Sg[i], transpose(Bg[i]))))
        push!(res, i=>T)
    end
    return res
end

function update_L(Sg, Bg, Tg)
    res = Dict{Int, Array}()
    g = length(Bg)
    for i = 1:g
        L = *(Sg[i], transpose(Bg[i]), inv(Tg[i]))
        push!(res, i=>L)
    end
    return res
end

function update_psi(p, Sg, Lg, Bg)
    res = Dict{Int, Float64}()
    g = length(Bg)
    for i = 1:g
        psi = *( /(1, p), tr( -(Sg[i], *(Lg[i], Bg[i], Sg[i]) ) ) )
        push!(res, i=>psi)
    end
    return res
end

function update_zmat(xmat, mug, Lg, psig, pig)
    N,p = size(xmat)
    g = length(Lg)
    res = Matrix{Float64}(undef, N, g)
    Ip = Matrix{Float64}(I, p, p)
    for i = 1:g
        pI = *(psig[i], Ip)
        mu = mug[:, i]
        cov = +( *( Lg[i], transpose(Lg[i]) ), pI)
        pi_den = *(pig[i], pdf(MvNormal(mu, cov), transpose(xmat)))
        res[:,i] = pi_den
    end
    return ./(res, sum(res, dims = 2))
end

function mapz(Z)
    N,g = size(Z)
    res = Vector{Int}(undef, N)
    for i = 1:N
        res[i] = findmax(Z[i,:])[2]
    end
    return res
end

function mppca_ll(N, p, pig, q, psig, Lg, Sg)
    g = length(Lg)
    l1,l2,l3 = (0,0,0)
```

```
    c1 = /(N,2)
    c2 = *(-1, c1, p, g, log( *(2,pi) ))
    for i = 1:g
        l1 += log(pig[i])
        l2 += log(wbidet(q, p, psig[i], Lg[i]))
        l3 += tr(*(Sg[i], wbiinv(q, p, psig[i], Lg[i])))
      end
      l1b = *(N, l1)
      l2b = *(-1, c1, l2)
      l3b = *(-1, c1, l3)
      return(+(c2, l1b, l2b, l3b))
end

function mppca_fparams(p, q, g)
    ## number of free parameters in the ppca model
    c1 = *(p, q)
    c2 = *(-1, q, -(q, 1), 0.5)
    return(+( *( +(c1, c2), g), g))
end

function mppca_proj(X, G, map, L)
    res = Dict{Int, Array}()
    for i in 1:G
        coef = svd(L[i]).U
        sel = map .== i
        proj = *(X[sel, :], coef)
        push!(res, i=>proj)
    end
    return(res)
end
```

Having created helper functions in the previous code block, an AECM algorithm for the MPPCA model is implemented in the following code block. The values of $\hat{z}_{ig}$ can be initialized in a number of ways, e.g., randomly or using the results of k-means clustering.

```
using Clustering
include("chp6_ppca_functions.jl")
include("chp6_mixppca_functions.jl")

## MPPCA function
function mixppca(X; q = 2, G = 2, thresh = 1e-5, maxit = 1e5, init = 1))
    ## initializations
    N, p = size(X)
    ## z_ig
    if init == 1
      ## random
      zmat = rand(Uniform(0,1), N, G)
      ## row sum to 1
      zmat = ./(zmat, sum(zmat, dims = 2))
    elseif init == 2
      # k-means
      kma = kmeans(permutedims(X), G; init=:rand).assignments
      zmat = zeros(N,G)
      for i in 1:N
        zmat[i, kma[i]] = 1
```

```
      end
   end
   n_g = sum(zmat, dims = 1)
   pi_g = /(n_g, N)
   mu_g = muG(G, X, zmat)
   S_g = sigmaG(mu_g, X, zmat)
   L_g = init_LambdaG(q, G, S_g)
   psi_g = init_psi(G, S_g, L_g)
   B_g = init_dict0(G, q, p)
   T_g = init_dict0(G, q, q)

   conv = true
   iter = 1
   ll = zeros(100)
   ll[1] = 1e4

   # while not converged
   while(conv)

     ## update pi_g and mu_g
     n_g = sum(zmat, dims = 1)
     pi_g = /(n_g, N)
     mu_g = muG(G, X, zmat)
     if iter > 1
       ## update z_ig
       zmat = update_zmat(X, mu_g, L_g, psi_g, pi_g)
     end
     ## compute S_g, Beta_g, Theta_g
     S_g = sigmaG(mu_g, X, zmat)
     B_g = update_B(q, p, psi_g, L_g)
     T_g = update_T(q, B_g, L_g, S_g)

     ## update Lambda_g psi_g
     L_g_new = update_L(S_g, B_g, T_g)
     psi_g_new = update_psi(p, S_g, L_g, B_g)
     ## update z_ig
     zmat = update_zmat(X, mu_g, L_g_new, psi_g_new, pi_g)

     iter += 1

     if iter > maxit
         conv = false
         println("mixppca() while loop went past maxit parameter. iter =
         $iter")
     else
       ## stores at most the 100 most recent ll values
       if iter <= 100
         ll[iter] = mppca_ll(N, p, n_g, pi_g, q, psi_g, L_g, S_g)
       else
         ll = circshift(ll, -1)
         ll[100] = mppca_ll(N, p, n_g, pi_g, q, psi_g, L_g, S_g)
       end
       if 2 < iter < 101
         ## scales the threshold by the latest ll value
         thresh2 = *(-1, ll[iter], thresh)
         if aitkenaccel(ll[iter], ll[(iter-1)], ll[(iter-2)], thresh2)
           conv = false
         end
       else
         thresh2 = *(-1, ll[100], thresh)
         if aitkenaccel(ll[100], ll[(99)], ll[(98)], thresh2)
           conv = false
```

```
          end
        end
      end ## if maxit
      L_g = L_g_new
      psi_g = psi_g_new
    end ## while

    map_res = mapz(zmat)
    proj_res = mppca_proj(X, G, map_res, L_g)

    if iter <= 100
      resize!(ll, iter)
    end
    fp = mppca_fparams(p, q, G)

    bic_res = BiC(ll[end], N, q, fp)

    return Dict(
      "iter" => iter,
      "ll" => ll,
      "beta" => B_g,
      "theta" => T_g,
      "lambda" => L_g,
      "psi" => psi_g,
      "q" => q,
      "G" => G,
      "map" => map_res,
      "zmat" => zmat,
      "proj" => proj_res,
      "bic" => bic_res
    )
  end

mixppca_12k = mixppca(cof_mat_c, q=1, G=2, maxit = 1e6, thresh = 1e-3,
    init=2)
```

Predicted Classifications

Predicted classifications are given by the values of (6.23) after the AECM algorithm has converged. These predicted classifications are inherently soft, i.e., $\hat{z}_{ig} \in [0, 1]$; however, in many practical applications, they are hardened by computing maximum *a posteriori* (MAP) classifications:

$$\text{MAP}\{\hat{z}_{ig}\} = \begin{cases} 1 & \text{if } g = \arg\max_h\{\hat{z}_{ih}\}, \\ 0 & \text{otherwise.} \end{cases}$$

6.6.3   Illustrative Example: Coffee Data

The MPPCA model is applied to the `coffee` data, using the code in the previous block. Models are initialized with k-means clustering and run for $G = 2, 3, 4$ and $q = 1, 2, 3$. Using the BIC, we choose a model with $G = 2$ groups (i.e., mixture components) and $q = 1$ latent component (Figure 6.10). This model correctly classifies all of the coffees into their respective varieties, i.e., the MAP classifications exactly match the varieties.

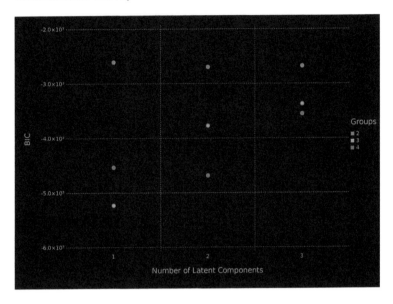

Figure 6.10 BIC values for the MPPCA models with different numbers of groups and latent components for the `coffee` data.

6.7 COMMENTS

Analogous to our comments in Section 5.8, it is worth emphasizing that only selected unsupervised learning techniques have been covered, along with Julia code needed for implementation. Just as Chapter 5 cannot be considered a thorough introduction to supervised learning, the present chapter cannot be considered a thorough introduction to unsupervised learning. However, the Julia code covered herein should prove most instructive. Furthermore, it is worth noting that two further unsupervised learning approaches — one of which can be considered an extension of MPPCA — are discussed in Chapter 7.

Two techniques for clustering, or unsupervised learning, have been considered herein. While there are a host of clustering techniques available (see, e.g., Everitt et al., 2011), the authors of the present monograph have a preference for mixture model-based approaches and a wide range of such approaches are available (see McNicholas, 2016a,b, for many examples). In the PPCA example, the BIC was used to select q and, when using mixtures thereof, the BIC was used to select both G and q. For completeness, it is worth noting that the BIC is a popular tool for mixture model selection in general (see McNicholas, 2016a, Section 2.4, for further details).

## CHAPTER 7

R Interoperability

T HE PRIMARY purpose of this chapter is to illustrate the
interoperability between R and Julia. Considering the wide
range of contributed libraries available for R, the ease with which R
can be called from Julia is a significant advantage. After the basics
are covered, two case studies are used for illustration. The first,
using the coffee data, also introduces two unsupervised learning
approaches: one that can be viewed in the context of a generalization of the MPPCA model and another that performs simulta-
neous dimension reduction and clustering. The second, using the
food data, illustrates random forests, a tree ensemble method that
can be regarded as an alternative to boosting.

7.1   ACCESSING R DATASETS

The RDatasets.jl package provides access to some of the most
commonly used R datasets. At the time of writing, there are 733 R
datasets available, coming from base R and some of the most popular R packages. The RDatasets.jl package can be thought of as
a port of the Rdatasets package in R. The RDatasets.jl package contains the RDatasets.datasets() function, which returns
a Julia DataFrame object listing the package, dataset, dataset title
and the number of rows and columns in the dataset. The columns
can be searched to find the package or dataset we are interested
in.

```
using RDatasets

## look for crabs in available datasets
rds = RDatasets.datasets()
```

```
filter(x -> occursin("crab", x[:Dataset]), rds)

crabs = dataset("MASS", "crabs")
print(crabs[1:5, :])
```

The RData.jl package allows users to read .Rdata and .rda files into Julia. While not all R types can (currently) be converted into Julia types, the major R types are well supported at present (Table 7.1).

TABLE 7.1   Major R types and their Julia equivalent.

R	Julia
vector	VectorType
factor	CategoricalArray
data.frame	DataFrame

The RData.jl package uses the load() function to read in R data files. The load() function returns a Julia dictionary of the form Dict{String, Any}, where the key is the R dataset name and the value is the Julia DataFrame representing the R data. Users can read in .Rdata and .rda files they have created or accessed from their favourite R packages and read them into Julia. In the below example, we read in the wine and coffee datasets from the pgmm package in R and convert them into Julia DataFrame objects.

```
using RData

# Read in wine data (in two steps)
wine = RData.load("wine.rda")
wine_df = wine["wine"]
println("wine_df: ", typeof(wine_df))

# Read in coffee data (in two steps)
coffee_df = RData.load("coffee.rda")["coffee"]
print("coffee_df[1:5,:]:\n", coffee_df[1:5,:])
```

7.2   INTERACTING WITH R

The RCall.jl package allows Julia users to interact directly with R packages. The RCall.jl package is written in Julia making it very easy to use. Upon installation, it checks for an existing R installation and, if one does not exist, it will install the version of R distributed by anaconda (anaconda.org/r/r-base). After updating a system's R environment, it is recommended to rebuild the RCall package so it picks up the updated version of R.

```
## Julia 0.6.x
Pkg.build("RCall")

## Julia 1.0.x
] build RCall
```

There are four ways to use RCall to interact with the R installation:

1. R REPL mode.

2. The @rput and @rget macros.

3. The R" " string macro.

4. RCall API: recal(), rcall(), rcopy().

Methods 1 and 2 above allow the user to interact with the R session from the REPL environment. Herein, we will focus on methods 3 and 4 because they work well inside larger Julia programs and we find that they have productivity advantages for a practicing data scientist.

RCall package uses $ for variable substitution, i.e., it sends the Julia object prepended with $ to the R environment for evaluation. In the following code block, R"chisq.test($x)" sends the Julia object x to R. Then R will run a chi-square test on x (as if it were a contingency table with one row) and return the result to the Julia environment. The result returned to Julia is of type RObject, which is a Julia wrapper for the R object.

```
using RCall

# Perform a Chi squared test using R fundtion chisq.test()
x = Int[10,39,50,24]
R"chisq.test($x)"

# RObject{VecSxp}
#
#      Chi-squared test for given probabilities
#
# data:  `#JL`$x
# X-squared = 29.748, df = 3, p-value = 1.559e-06
```

Some caution is required when using RCall. For example, if the expression representing the substitution is a valid R command or syntax, the results can be unexpected. Consider the following code block. In the first example, the Julia array index is actually a column in the `crabs` dataset, and the submitted command returns the data in the index column of the R dataset. The second example, however, is evaluated as expected in R, i.e., the rows of the `crabs` dataframe corresponding to the integers in the index array are returned.

```
## Here, the Julia variable is not used
## The data in the  index field is returned
index = [1,2,5]
print(R"MASS::crabs$index")
# RCall.RObject{RCall.IntSxp}
#  [1]  1  2  3  4  ..

## Here, the Julia variable is used
R"MASS::crabs[$index, ]"
# RCall.RObject{RCall.VecSxp}
#     sp sex index FL  RW   CL   CW  BD
# 1  B   M     1 8.1 6.7 16.1 19.0 7.0
# 2  B   M     2 8.8 7.7 18.1 20.8 7.4
# 5  B   M     5 9.8 8.0 20.3 23.0 8.2
```

An alternative is to use the string macro, which allows large blocks of code to be submitted to R when they are enclosed in triple quotes, i.e., `"""` . . . `"""`. Consider the following code block, which simulates data for a logistic regression in Julia. This code uses the `Distributions.jl` package to generate random numbers from a Bernoulli distribution, where the probabilities for the Bernoulli distribution come from applying the `inv-logit` function to the linear function `lf`. Note that the logit of $p \in (0, 1)$ is given by

$$\text{logit}(p) = \log\left(\frac{p}{1-p}\right),$$

and the inverse-logit, or logistic function, of $q \in \mathbb{R}$ is

$$\text{logit}^{-1}(q) = \frac{1}{1 + \exp\{-q\}}.$$

The logit has an important interpretation in statistics and data science: if an event E has probability p, then

$$\log \text{odds}(E) = \text{logit}(p).$$

Returning to the following code block, notice that the variables `x1`, `x2` and `y` are sent to R. In R, these three variables, along with a third predictor variable, are made into a dataframe. The dataframe is then used as input to the `glm()` function in R, which performs a logistic regression. The results of the logistic regression are then returned to Julia. The results are as one would expect: x1 and x2 have significant regression coefficients because of their relationship to the binary outcome through the linear predictor, but x3 is not related to the outcome — it is generated as noise from a normal distribution — and does not have a statistically significant relationship to the outcome.

```julia
using RCall, Distributions, StatsBase, Random
Random.seed!(636)

# Simulate data for logistic regression
N = 1000
x1 = randn(N)
x2 = randn(N)
x0 = fill(1.0, N)

# A linear function of x1 and x2
lf = x0 + 0.5*x1 + 2*x2

# The inv-logit function of lf
prob = 1 ./ ( x0+ exp.(-lf))

# Generate y
y = zeros(Int64, N)
for i = 1:N
  y[i] = rand(Binomial(1, prob[i]), 1)[1]
end

# Run code in R
# Note that x3 is generated in R (randomly from a normal distribution) but
# x1 and x2 are sent from Julia to R
R"""
set.seed(39)
n <- length($x1)
df <- data.frame(x.1 = $x1, x.2 = $x2, y = $y, x.3 = rnorm(n))
fit1 <- glm(y ~ x.1 + x.2 + x.3, data = df, family = "binomial")
summary(fit1)
"""
# Coefficients:
#               Estimate Std. Error z value Pr(>|z|)
# (Intercept)   0.86586    0.09077    9.540  < 2e-16 ***
# x.1           0.46455    0.08640    5.377 7.57e-08 ***
# x.2           1.86382    0.12481   14.934  < 2e-16 ***
# x.3           0.14940    0.08521    1.753   0.0795 .


# Odds ratios for x1 and x2, respectively
exp(0.46455)
# 1.592
exp(1.86382)
# 6.448
```

The `RCall` API has three commonly used functions: `rcopy()`, `reval()`, and `rcall()`. The `rcopy()` function converts R objects into Julia objects, and the Julia type for a given R object is determined by some heuristic criterion. Some conversion examples and their types are given in the following code block:

```
# Heuristic conversion examples
d1 = rcopy(R"""data.frame(v1 = 1:2, v2=c("Data", "Science"))""")
println("type d1: ", typeof(d1))
# type d1: DataFrames.DataFrame

l1 = rcopy(R"list(2.3, 'red') ")
println("type l1: ", typeof(l1))
# type l1: Array{Any,1}

l2 = rcopy(R"list(v1=2.3, v2='red')")
println("type l2: ", typeof(l2))
# type l2: DataStructures.OrderedDict{Symbol,Any}

# Note that rcopy() will force an exact conversion if the type is
# specified as the first argument
l3 = rcopy(Array{String,1}, R"""c("Data","Science")""")
println("type l3: ", typeof(l3))
# type l3: Array{String,1}
```

The `reval()` function takes a Julia string as input and evaluates the string in the R environment as if it were R code. If the string is accepted by R as a valid input, `reval()` returns an RObject object back to the Julia environment. Note that, if $ is used in the string representing the R code, it must be Escaped, i.e., "\$". The `rcall()` function is used to make R function calls. If a function call is successfully evaluated in R, then the result is returned as an RObject. The first argument to `rcall()` is the function being called and it is specified as a Julia symbol type `:R_function_name`. Additional arguments to the R function can be specified after its name; the arguments are separated by commas and they must exist in the Julia environment. The following code block illustrates the use of `reval()` and `rcall()`. Note that the `lm()` function in R fits a linear model.

```
# Use reval() to pull the simulated data into Julia
# df is the R dataframe used to produce the glm object fit1
# df contains the data we simulated in Julia and sent to R via the R""
# string macro

df_r = reval("df")

# Then, use rcall() to run the lm() function in R
# note that df_r is in the Julia environment and is being passed back to R
```

```
lm_r = rcall(:lm, "x.1 ~ y", df_r)
# or
lm_r = rcall(:lm, "as.formula('x.1~y')", df_r)

# Rather than using reval() followed by rcopy(), they can be used
# in conjuction
lm_df = rcopy(reval("summary(lm(x.1 ~ y, df))\$coefficients"))

print("typeof(lm_df): $(typeof(lm_df))")
# typeof(lm_df): Array{Float64,2}
```

The rcall() function requires some extra syntax when the user wishes to pass keyword arguments that have "." in their names or the names of R objects already initialized in the R environment. The former case is a problem because Julia does not allow "." in its variable names. The @var_str(str) macro, distributed with the RCall.jl package was written to remedy this. It allows the keyword name containing the "." to be passed from Julia to R. The latter case can be solved by using the Symbol constructor to create a new symbol from the R object name, allowing the R function to recognize it as an R object in its environment. The following code block uses reval() to create a custom R dataframe in the R environment. The rcall() is required to use the Symbol constructor to access this new dataframe and the var_str macro to pass the "scale." keyword to the prcomp() function in R.

```
## create data in R
reval("""data(crabs, package = "MASS")
  df.pca <- subset(crabs, select = -c(sp, sex, index))""")

## rcall() using both the var_str macro and Symbol constructor
prcomp_r = rcall(:prcomp, Symbol("df.pca"), center = true,
  var"scale." = true)
```

7.3 EXAMPLE: CLUSTERING AND DATA REDUCTION FOR THE COFFEE DATA

7.3.1 Coffee Data

Recall that the coffee data contain 43 samples of the Arabica and Robusta species with 12 of the chemical constituents available for each sample. As mentioned in Section 1.5.3, an interesting feature of these data is that two variables, fat and caffeine, perfectly separate the Arabica and Robusta classes. Therefore, it is interesting to consider whether a clustering technique can be used that relies heavily on dimension reduction — after all, only two of the 12 variables should be needed.

7.3.2 PGMM Analysis

PGMM Family

Recall the MPPCA model (Section 6.6), where we have a Gaussian mixture model with covariance matrix $\Sigma_g = \Lambda_g\Lambda'_g + \psi_g\mathbf{I}_p$. Around the same time as the MPPCA model was introduced, Ghahramani and Hinton (1997) developed a related mixture of factor analyzers model with $\Sigma_g = \Lambda_g\Lambda'_g + \Psi$. Shortly thereafter, McLachlan and Peel (2000) introduced a more general mixture of factor analyzers model with $\Sigma_g = \Lambda_g\Lambda'_g + \Psi_g$. McNicholas and Murphy (2008) develop a family of eight parsimonious Gaussian mixture models (PG-MMs) for clustering by imposing, or not, each of the constraints $\Lambda_g = \Lambda$, $\Psi_g = \Psi$, and $\Psi_g = \psi_g\mathbf{I}_p$. Members of the PGMM family have between $pq - q(q-1)/2 + 1$ and $G[pq - q(q-1)/2] + Gp$ free parameters in the component covariance matrices (see Table 7.2). Note that the MPPCA model is called the UUC model in the PGMM family nomenclature (Table 7.2).

Table 7.2 The nomenclature and covariance structure for each member of the PGMM family of McNicholas and Murphy (2008), where "C" denotes "constrained", i.e., the constraint is imposed, and "U" denotes "unconstrained", i.e., the constraint is not imposed.

$\Lambda_g = \Lambda$	$\Psi_g = \Psi$	$\Psi_g = \psi_g\mathbf{I}_p$	Σ_g
C	C	C	$\Lambda\Lambda' + \psi\mathbf{I}_p$
C	C	U	$\Lambda\Lambda' + \Psi$
C	U	C	$\Lambda\Lambda' + \psi_g\mathbf{I}_p$
C	U	U	$\Lambda\Lambda' + \Psi_g$
U	C	C	$\Lambda_g\Lambda'_g + \psi\mathbf{I}_p$
U	C	U	$\Lambda_g\Lambda'_g + \Psi$
U	U	C	$\Lambda_g\Lambda'_g + \psi_g\mathbf{I}_p$
U	U	U	$\Lambda_g\Lambda'_g + \Psi_g$

McNicholas and Murphy (2010) further parameterize the mixture of factor analyzers component covariance structure by writing

$$\Psi_g = \omega_g\Delta_g,$$

where $\omega_g \in \mathbb{R}^+$ and $\mathbf{\Delta}_g$ is a diagonal matrix with $|\mathbf{\Delta}_g| = 1$. The resulting mixture of modified factor analyzers model has component covariance structure

$$\mathbf{\Sigma}_g = \mathbf{\Lambda}_g \mathbf{\Lambda}'_g + \omega_g \mathbf{\Delta}_g.$$

In addition to the constraint $\mathbf{\Lambda}_g = \mathbf{\Lambda}$, all legitimate combinations of the constraints $\omega_g = \omega$, $\mathbf{\Delta}_g = \mathbf{\Delta}$, and $\mathbf{\Delta}_g = \mathbf{I}_p$ are imposed, resulting in a family of 12 parsimonious Gaussian mixture models (Table 7.3). Hereafter, this family will be called the PGMM family.

Table 7.3 The covariance structure and nomenclature for each member of the PGMM family of McNicholas and Murphy (2010).

$\mathbf{\Lambda}_g = \mathbf{\Lambda}$	$\mathbf{\Delta}_g = \mathbf{\Delta}$	$\omega_g = \omega$	$\mathbf{\Delta}_g = \mathbf{I}_p$	$\mathbf{\Sigma}_g$
C	C	C	C	$\mathbf{\Lambda}\mathbf{\Lambda}' + \omega\mathbf{I}_p$
C	C	U	C	$\mathbf{\Lambda}\mathbf{\Lambda}' + \omega_g\mathbf{I}_p$
U	C	C	C	$\mathbf{\Lambda}_g\mathbf{\Lambda}'_g + \omega\mathbf{I}_p$
U	C	U	C	$\mathbf{\Lambda}_g\mathbf{\Lambda}'_g + \omega_g\mathbf{I}_p$
C	C	C	U	$\mathbf{\Lambda}\mathbf{\Lambda}' + \omega\mathbf{\Delta}$
C	C	U	U	$\mathbf{\Lambda}\mathbf{\Lambda}' + \omega_g\mathbf{\Delta}$
U	C	C	U	$\mathbf{\Lambda}_g\mathbf{\Lambda}'_g + \omega\mathbf{\Delta}$
U	C	U	U	$\mathbf{\Lambda}_g\mathbf{\Lambda}'_g + \omega_g\mathbf{\Delta}$
C	U	C	U	$\mathbf{\Lambda}\mathbf{\Lambda}' + \omega\mathbf{\Delta}_g$
C	U	U	U	$\mathbf{\Lambda}\mathbf{\Lambda}' + \omega_g\mathbf{\Delta}_g$
U	U	C	U	$\mathbf{\Lambda}_g\mathbf{\Lambda}'_g + \omega\mathbf{\Delta}_g$
U	U	U	U	$\mathbf{\Lambda}_g\mathbf{\Lambda}'_g + \omega_g\mathbf{\Delta}_g$

Note that eight of the 12 models in Table 7.3 are equivalent to models in Table 7.2 — the four models with no counterpart in Table 7.2 are CCUU, UCUU, CUCU, and UUCU. In other words, the models in Table 7.2 are a subset of those in Table 7.3. The pgmm package for R implements all 12 PGMM models for model-based clustering and classification. A key feature of the PGMM family is that all members have $\mathcal{O}(p)$ covariance parameters, i.e., the number of covariance parameters is linear in the dimensionality of the data. Clearly, this is very important in the analysis of high-dimensional data. However, it is also important in cases where many of the variables are not contributing, *per se*, to the model.

Analysis of Coffee Data

In the following code block, a PGMM analysis of the `coffee` dataset is carried out.

```
## Clustering the coffee data with PGMM
using DataFrames, RCall

## make a copy of the Julia coffee dataframe
## remove the Variety and Country columns before clustering
x = deepcopy(coffee_df)
delete!(x, [:Variety, :Country])
println("size(x): $(size(x))")
# size(x): (43, 12)

## Scale cols of x in R
x_scaled = rcall(:scale, x)

## load the pgmm library in R
reval("library(pgmm)")

## Run PGMM using pgmmEM() in R
## Note the parameter values are all defined in the Julia environment
pgmm_r = rcall(:pgmmEM, rG=2:3, rq=1:3, zstart=2, icl=true, x_scaled)

## change the RObject into a Julia dictionary
pgmm_j = rcopy(pgmm_r)
println("\n\ntypeof(pgmm_j): ", typeof(pgmm_j))
# typeof(pgmm_j): OrderedCollections.OrderedDict{Symbol,Any}

## print the dictionary keys
for (k,v) in pgmm_j
  println("pgmm_j: key: ", k)
end

##Make an new dataframe for the analysis
df_results = deepcopy(coffee_df[[:Variety, :Fat, :Caffine]])
df_results[:Map] = pgmm_j[:map]
println("names(df_results): $(names(df_results))")
# names(df_results): Symbol[:Variety, :Fat, :Caffine, :Map]

## classification table
ct1 = by(df_results, [:Map, :Variety], nrow)
ct1
# 2x3 DataFrame
# | Row | Map     | Variety | x1    |
# |     | Float64 | Float64 | Int64 |
# +-----+---------+---------+-------+
# | 1   | 1.0     | 1.0     | 36    |
# | 2   | 2.0     | 2.0     | 7     |
```

The results of this analysis reveal that a $q = 1$ factor, $G = 2$ component CCUU model is selected by the BIC. Similar to the MPPCA analysis (Section 6.6.3), this model obtains perfect clustering on the `coffee` data.

7.3.3 VSCC Analysis

VSCC

Andrews and McNicholas (2014) introduce the variable selection for clustering and classification (VSCC) technique. The goal of VSCC is to find a subset of variables that simultaneously minimizes the within-group variance and maximizes the between-group variance. In other words, VSCC finds variables that show separation between the desired groups. The within-group variance for the jth variable can be written

$$\mathcal{W}_j = \frac{1}{n} \sum_{g=1}^{G} \sum_{i=1}^{n} z_{ig}(x_{ij} - \mu_{gj})^2,$$

where x_{ij} is the value of the jth variable for the ith observation, μ_{gj} is the mean of the jth variable in the gth component, and n and z_{ig} have the usual meanings. The variance within the jth variable that is not accounted for by $\mathcal{W}_j$, i.e., $\sigma_j^2 - \mathcal{W}_j$, gives an indication of the variance between groups. In general, calculation of this residual variance is needed; however, if the data are standardized to have equal variance across variables, then a variable that minimizes the within-group variance will also maximize the leftover variance. Accordingly, Andrews and McNicholas (2014) describe the VSCC method in terms of variables that are standardized to have zero mean and unit variance. The VSCC approach, which also uses the correlation between variables, is described in detail in Andrews and McNicholas (2014) and in Chapter 4 of McNicholas (2016a).

Analysis of Coffee Data

```
## Clustering and (explicit) variable reduction for the coffee
## data with VSCC

# This builds on the previous code block and we expect certain objects
#  to be initialized, such as x_scaled

reval("library(vscc)")

## run vscc on the scaled data in R and copy the results into a Julia
## dictionary
vscc_j = rcopy(rcall(:vscc, x_scaled))

## Array containing the best data columns
## Fat and Free Acid were identified
```

```
vscc_ts = vscc_j[:topselected]

## MAP classifications from the best VSCC model
vscc_map = vscc_j[:bestmodel][:classification]
df_results[:vscc_MAP] = vscc_map

## classification table
ct2 = by(df_results, [:vscc_MAP, :Variety], nrow)
ct2
# 2x3 DataFrame
# | Row | vscc_MAP | Variety | x1    |
# |     | Float64  | Float64 | Int64 |
# +-----+----------+---------+-------+
# | 1   | 1.0      | 1.0     | 36    |
# | 2   | 2.0      | 2.0     | 7     |

## updating the analysis dataframe from pgmm with
## the missing variable selected by VSCC
df_results[:Free_Acid] = coffee_df[:Free_Acid]
```

The results reveal that two variables are chosen — free acid and fat — and the selected $G = 2$ component model gives perfect clustering performance (Figure 7.1). Although caffeine and fat have perhaps more commonly been highlighted as separating the varieties of coffee, free acid and fat also separate the classes perfectly (Figure 7.1).

7.4 EXAMPLE: FOOD DATA

7.4.1 Overview

As an additional illustration of using RCall.jl to interact with R, we re-analyze the food data. A random forest algorithm is used to build a supervised learning model to predict student GPA from the 196 diet and nutrition related predictor variables. The random forest learner is built using the same training and test data we used to build the boosting learners in Chapter 5.

7.4.2 Random Forests

Random forests were introduced in Section 5.6. The random forest algorithm we use here is implemented in the R package ranger (Wright and Ziegler, 2017). The ranger package is a newer implementation of random forests, optimized for speed and high-dimensional data. The ranger learner was trained with the help of the train() function in the caret package (Kuhn, 2017) — CARET is short for "Classification And REgression Training" and simplifies building and training predictive learners in R. At the time

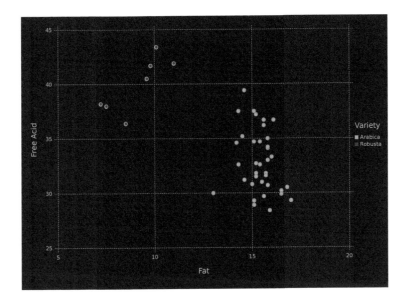

Figure 7.1   Scatterplot of the variables selected by `vscc` for the `coffee` data, coloured by the predicted classes (which are the same as the true classes).

of this writing, `caret` gives R users a unified interface to train and build 238 different predictive learners. Along with these learners, users can choose from a number of different re-sampling schemes and can enter custom evaluation metrics, as we did in Chapter 5 when we used MAE to train the boosting learner. The capabilities of `caret` are vast and are nicely illustrated by Kuhn and Johnson (2013).

We start the analysis by submitting the R code contained in `chp7_ranger.R` to the R environment. When the code in the file is run, it starts by loading the `caret` and `ranger` packages and setting the random number seed. It then loads some custom `caret` functions and parameter objects into the R environment. We will use these objects with `rcall()` and `train()` to train our random forest learners. The objects include some custom `trainControl` list objects, controlling how the cross-validation will be done. There is a function to calculate the MAE, a summary function that computes performance metrics across all the training re-samples, and a dataframe that represents the `ranger` parameter grid used to train the random forest learners. The `ranger` learners will be trained on different values of the number of variables to use at each split, the minimum node size and two different regression split rules.

From here, we load the food training and test data into the Julia environment. We are required to re-format it so it will pass through the `RCall` interface in the correct format for the `train()` function. As illustrated in the following code block, each set of data is separated into predictor and outcome arrays. These arrays are then converted into Julia dataframes.

```
reval("""source("./chp7_ranger.R")""")

## Prep data for R
## training data
y_tr = convert(Array{Float64}, df_train[:gpa])

sel_var = setdiff(names(df_train), [:gpa])
tmp = convert(Array, df_train[sel_var] )
x_tr =convert(Array{Float64}, collect(Missings.replace(tmp, 0)))
x_tr_df = convert(DataFrame, x_tr)
names!(x_tr_df, sel_var)

## testing data
y_tst = convert(Array{Float64}, df_test[:gpa])

sel_var = setdiff(names(df_test), [:gpa])
tmp = convert(Array, df_test[sel_var] )
x_tst =convert(Array{Float64}, collect(Missings.replace(tmp, 0)))
x_tst_df = convert(DataFrame, x_tst)
names!(x_tst_df, sel_var)
```

The first `ranger` learner is trained using 5-fold cross-validation. We grow an ensemble of 1000 trees on 16 different combinations of the three training parameters. The Julia code is given in the following code block. Because the `caret tuneGrid` and `trControl` objects are already defined in the R environment, we use the `Symbol` constructor to pass their names to the R environment. The `ranger()` function requires the number of trees to be specified by the `num.trees` function argument. To correctly pass this argument, it must be passed to the `@var_str` macro. We find it is more efficient to wrap the `rcall()` function call directly inside the `rcopy()` function. Given that we know `train()` returns an R list of results and `rcopy()` will convert this list to a Julia

`DataStructures.OrderedDict{Symbol,Any}`

object, we can append the `rcopy()` function call with the dictionary key corresponding to the `train()` result we are interested in.

```
## train the ranger model using 5-fold CV and 1000 trees
rf_tr_j1 = rcopy(
  rcall(:train, x_tr_df, y_tr , method = "ranger",
        trControl = Symbol("trainParam"),
        tuneGrid = Symbol("rfParam"),
        var"num.trees" = 500)
  )[:results]

## add the standard errors for the CV error
## used in the plots
rf_tr_j1[:MedAE_se] = map(x -> x / sqrt(5), rf_tr_j1[:MedAESD])
rf_tr_j1[:MedAE_min] = map(-, rf_tr_j1[:MedAE], rf_tr_j1[:MedAE_se])
rf_tr_j1[:MedAE_max] = map(+, rf_tr_j1[:MedAE], rf_tr_j1[:MedAE_se])

min_medae = minimum(rf_tr_j1[:MedAE])
min_err = filter(row -> row[:MedAE] == min_medae ,rf_tr_j1)

## apply the 1 SE method
one_se = min_err[1, :MedAE_max]
possible_models = filter(row -> row[:MedAE] <= one_se, rf_tr_j1)
```

The 5-fold cross-validation results are illustrated in Figure 7.2. The horizontal line represents one standard error above the smallest MAE observed across the 16 learners. These results are ambiguous in terms of which learner parametrization is likely to produce the best predictions. All but three estimates are below or touching the horizontal line. Applying the one-standard error method, these three candidate learner configurations would be excluded. Learners that could sample from more variables to split their tree nodes produced the smallest error estimates.

In the hopes of distinguishing a small handful of top performing parameterizations, we retrain the ranger learner using 10-fold cross-validation and a 5000-tree ensemble. The additional folds will tighten up the error bars and further trees could help the accuracy of the prediction estimates. The Julia code to do this is in the ensuing code block. The new cross-validation specifications are defined in the **trainParam10** object in the R environment.

```
## train the ranger model using 10-fold CV and 5000 trees
rf_tr_j2 = rcopy(
  rcall(:train, x_tr_df, y_tr , method = "ranger",
        trControl = Symbol("trainParam10"),
        tuneGrid = Symbol("rfParam"),
        var"num.trees" = 5000)
  )[:results]

## add the standard errors for the CV error
## used in the plots
rf_tr_j2[:MedAE_se] = map(x -> x / sqrt(5), rf_tr_j2[:MedAESD])
rf_tr_j2[:MedAE_min] = map(-, rf_tr_j2[:MedAE], rf_tr_j2[:MedAE_se])
rf_tr_j2[:MedAE_max] = map(+, rf_tr_j2[:MedAE], rf_tr_j2[:MedAE_se])
```

```
min_medae = minimum(rf_tr_j2[:MedAE])
min_err = filter(row -> row[:MedAE] == min_medae ,rf_tr_j2)

## apply the 1 SE method
one_se = min_err[1, :MedAE_max]
possible_models = filter(row -> row[:MedAE] <= one_se, rf_tr_j2)
```

The results are displayed in Figure 7.3. The additional cross-validation folds resulted in tighter error bars but no real separation in the different learner parameterizations. The dashed line indicates the cut-off for the one-standard error method. All the estimates are below the line, making choosing between the learner configurations more difficult. These results are not unexpected. It is well known that trying to tune the performance of random forests learners typically leads to only mild performance improvements (Kuhn and Johnson, 2013). The results of the default configuration is given by the right-most variance error bar in Figure 7.3, which the best performing configuration marginally improves upon. In our case, the smallest MAE improved from 0.309 to 0.278 but we are not able to identify a small number of preferred learner con-

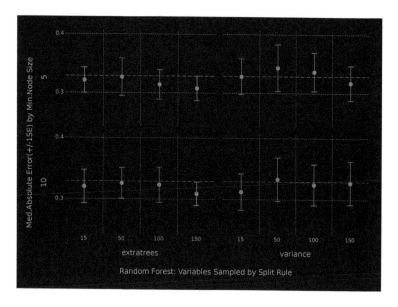

Figure 7.2 5-fold cross-validation results for the random forest learner trained on the food data.

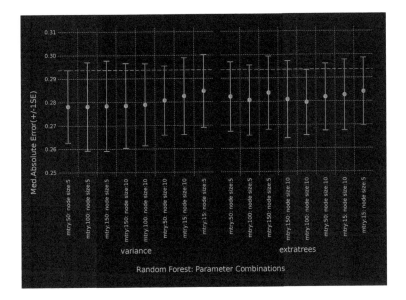

Figure 7.3 10-fold cross-validation results for the random forest learner trained on the `food` data.

figurations. This is in sharp contrast to the boosting results in Chapter 5 where, despite using 5-fold cross-validation, we could select a few combinations of learner parameters that performed better than the rest.

Given the ambiguous training results, we chose the simplest learner to predict the test data. This learner has a minimum node size of 10, making for trees of smaller depth, randomly chooses 15 variables at each split in the trees and uses the `extratrees` splitting criteria to build the trees (Geurts et al., 2006). The `extratrees` criteria, which is a parameter in `ranger`, chooses cutpoints from the candidate variables at random. When called individually and not through `train()`, `ranger` and its `predict` function require the predictor and outcome data to be in one dataframe. We call these Julia dataframes, `ranger_tr` and `ranger_tst`, respectively. An empty Julia dataframe is created to store the test set results. It has a field for each of the three parameters and the two error metrics. We loop over the parameter settings, using `rcall()` and `rcopy()` to make the R function calls and return the predictions to the Julia environment. The predictions are used to calculate the error metrics, MAE and RMSE, which are stored in an array along with the learner parameter values associated with them. This array is then added to the testing result dataframe via the **push!**() function.

```
## Test set performance

## ranger predict() needs one DF with outcome and predictors
ranger_tr = deepcopy(x_tr_df)
ranger_tr[:gpa] = y_tr
ranger_tst = deepcopy(x_tst_df)
ranger_tst[:gpa] = y_tst

## empty df to hold results
tst_results = DataFrame(
  mtry = Float64[],
  min_node_size = Float64[],
  splitrule = String[],
  medae = Float64[],
  rmse = Float64[])

## call ranger and make the predictions
tmp_ranger = rcall(:ranger, "gpa ~ .", data = ranger_tr, mtry= 15,
    var"num.trees" = 5000, var"min.node.size" = 10,
    splitrule = "extratrees")
tmp_pred = rcopy(rcall(:predict, tmp_ranger, ranger_tst))[:predictions]

## consolidate the results
tmp_array = [15, 10, "extratrees",
            medae(tmp_pred, y_tst), rmse(tmp_pred, y_tst)]
push!(tst_results, tmp_array)
```

When evaluated on the test set, the chosen learner produced an MAE of 0.263 and a RMSE of 0.431. These results are very competitive with the test set performance we saw from XGBoost in Chapter 5. It has been shown empirically that the random forest family of classifiers can outperform other well-known alternatives over a wide range of datasets (Fernández-Delgado et al., 2014). Given this, and their ease of training, random forests should be considered as an option when faced with a supervised learning problem.

The final step in this analysis is to calculate the variable importance measures on the overall data. We start by making a dataframe that contains the full dataset. The random forest learner is run on this data with the help of `rcall()` and the `@var_str` macro. The ranger learner is parameterized with the values identified using the one standard error method. The call to `ranger()` explicitly tells it to calculate the variable importance measures, as they are computationally expensive and not done by default. These variable importance measures are calculated using the `impurity`

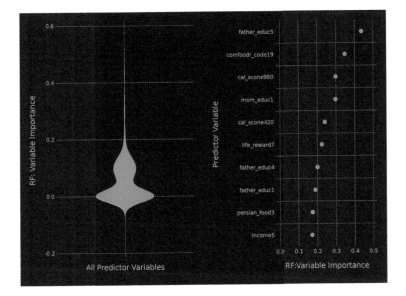

Figure 7.4 Random forest variable importance results for the food data.

variable importance mode in **ranger()**, which for regression is the variance of the responses. The code is detailed below and the variable importance results are plotted in Figure 7.4.

```
## Overall data
df_oa = append!(ranger_tr, ranger_tst)

## best test set parameterization
best_ranger_r = rcall(:ranger, "gpa ~ .", data = df_oa, mtry= 15,
  var"num.trees" = 5000, var"min.node.size" = 10,
  splitrule = "extratrees", importance = "impurity")

## Variable Importance
br_imp_r = rcall(:importance, best_ranger_r)

## rcopy does not preserve the names
br_imp_j = DataFrame(varname =  map(x -> string(x), names(br_imp_r)),
    vi = rcopy(br_imp_r) )

## sort by variable importance
sort!(br_imp_j, :vi, rev=true)
```

The violin plot on the left-hand side of Figure 7.4 indicates that the majority of the predictor variables are not important in the prediction of students' GPA. In the context of the y-axis, we can see that almost all of the predictor variables have importance values below 0.2, with a median value close to 0 (it is 0.002). The results advance the idea that the following characteristics are important in predicting a student's GPA: father's education, favourite comfort foods (or lack thereof), accurately estimating calories in food, and family income.

Julia and R Packages Used Herein

The R packages used herein are detailed in Table A.1, and the Julia packages used herein are listed in Table A.2.

Table A.1 The R packages used herein, with version number and relevant citations.

Name	Version	Relevant Citations
caret	6.0-78	Kuhn (2017), Kuhn and Johnson (2013)
pgmm	1.2.2	McNicholas et al. (2018), McNicholas (2010), McNicholas and Murphy (2008, 2010)
ranger	0.9	Wright and Ziegler (2017)
vscc	0.2	Andrews and McNicholas (2013, 2014)

Table A.2 The Julia packages used herein, with version number and URL.

Name	Version	URL
CSV.jl	0.4.2	github.com/JuliaData/ CSV.jl
Cairo.jl	0.5.6	github.com/JuliaGraphics/ Cairo.jl
Clustering.jl	0.12.1	github.com/JuliaStats/ Clustering.jl
D3DecisionTrees.jl	0.0.0	github.com/ValdarT/ D3DecisionTrees.jl
D3Trees.jl	0.3.0	github.com/sisl/ D3Trees.jl
DataValues.jl	0.4.5	github.com/queryverse/ DataValues.jl
DataFrames.jl	0.14.1	github.com/JuliaData/ DataFrames.jl
DecisionTree.jl	0.8.1	github.com/JuliaData/ DecisionTree.jl
Distances.jl	0.7.3	github.com/JuliaStats/ Distances.jl
Distributions.jl	0.16.4	github.com/JuliaStats/ Distributions.jl
GLM.jl	1.0.1	github.com/JuliaStats/ GLM.jl
Gadfly.jl	1.0.0	github.com/GiovineItalia/ Gadfly.jl
JLD2.jl	0.1.2	github.com/JuliaIO/ JLD2.jl
MLBase.jl	0.8.0	github.com/JuliaStats/ MLBase.jl
MLDataUtils.jl	0.4.0	github.com/JuliaML/ MLDataUtils.jl
MLLabelUtils.jl	0.4.1	github.com/JuliaML/ MLLabelUtils.jl
Query.jl	0.10.1	github.com/queryverse/ Query.jl
RCall.jl	0.12.1	github.com/JuliaInterop/ RCall.jl
RData.jl	0.5.0	github.com/JuliaData/ RData.jl
RDatasets.jl	0.6.1	github.com/johnmyleswhite/ RDatasets.jl
StatsBase.jl.jl	0.25.0	github.com/JuliaStats/ StatsBase.jl
StatsModels.jl.jl	0.3.1	github.com/JuliaStats/ StatsModels.jl
XGBoost.jl	0.2.0	github.com/dmlc/ XGBoost.jl

Variables for Food Data

Variable names for the food data are given in Tables B.1.

Table B.1 Variables for the beer dataset.

Variable	Description
GPA	Grade point average.
Gender	Female (1); male (2).
breakfast	What do you associate with breakfast? Cereal (1); doughnut (2).
calories_chicken	Guess for calories in chicken piadina: 265 (1), 430 (2), 610 (3), 720 (4).
calories_day	Importance of daily amount of calorie consumption: I don't know how many calories I should consume (1); not at all important (2); moderately important (3); very important (4).
calories_scone	Guess for calories in a scone.
coffee	What do you associate with coffee? Creamy frappuccino (1); espresso (2).
comfort_food	List 3–5 comfort foods.
comfort_food_reasons	List up to three reasons you consume comfort food.
comfort_food_reasons_coded	Stress (1); boredom (2); depression/sadness (3); hunger (4); laziness (5); cold weather (6); happiness (7); watching TV (8); none (9).

Table B.1 Variables for the beer dataset (continued).

Variable	Description
cook	How often do you cook? Every day (1); a couple of times a week (2); whenever I can, but that is not very often (3); I only help a little during holidays (4); never, I really do not know my way around a kitchen (5).
cuisine	What type of cuisine did you eat growing up? American (1); Mexican/Spanish (2); Korean/Asian (3); Indian (4); American inspired international dishes (5); other (6).
diet_current	Describe your current diet.
diet_current_coded	Healthy/balanced/moderated (1); unhealthy/cheap/too much/random (2); the same thing over and over (3); unclear (4).
drink	Binary: What do you associate with drink? Orange juice (1); soda (2).
eating_changes	Eating changes since admitted to college.
eating_changes_coded	Worse (1); better (2); the same (3); unclear (4).
eating_changes_coded1	Eat faster (1); bigger quantity (2); worse quality (3); same food (4); healthier (5); unclear (6); drink coffee (7); less food (8); more sweets (9); timing (10); more carbs or snacking (11); drink more water (12); more variety (13).
eating_out	Frequency of eating out in a typical week: Never (1); 1–2 times (2); 2–3 times (3); 3–5 times (4); every day (5).
employment	Do you work? Yes, full time (1); yes, part time (2); no (3); other (4).
ethnic_food	How likely are you to eat ethnic food? Very unlikely (1); unlikely (2); neutral (3); likely (4); very likely (5).
exercise	How often do you exercise in a regular week? Every day (1); twice or three times per week (2); once a week (3); sometimes (4); never (5).

Table B.1 Variables for the beer dataset (continued).

Variable	Description
father_education	Less than high school (1); high school degree (2); some college degree (3); college degree (4); graduate degree (5).
father_profession	Profession of father.
fav_cuisine	Favourite cuisine.
fav_cuisine_coded	None (0); Italian/French/Greek (1); Spanish/Mexican (2); Arabic/Turkish (3); Asian/Chinese/Thai/Nepal (4); American (5); African (6); Jamaican (7); Indian (8).
fav_food	Favourite food was: cooked at home (1); store bought (2); both bought at store and cooked at home (3).
food_childhood	Favourite childhood food.
fries	Which do you associate with fries? McDonald's fries (1); home fries.
fruit_day	How likely are you to eat fruit in a regular day? Very unlikely (1); unlikely (2); neutral (3); likely (4); very likely (5).
grade_level	Freshman (1), Sophomore (2), Junior (3), Senior (4).
greek_food	How likely are you to eat Greek food when available? Very unlikely (1); unlikely (2); neutral (3); likely (4); very likely (5).
healthy_feeling	"I feel very healthy!": Strongly agree (1) to strongly disagree (10).
healthy_meal	What is a healthy meal?
ideal_diet	Describe your ideal diet.
ideal_diet_coded	Portion control (1); adding veggies/eating healthier food/adding fruit(2); balance (3); less sugar (4); home cooked/organic (5); current diet (6); more protein (7); unclear (8).
income	<$15,000 (1); $15,001–$30,000 (2); $30,001–$50,000 (3); $50,001–$70,000 (4); $70,001–$100,000 (5); >$100,000 (6).
indian_food	How likely are you to eat Indian food when available? Very unlikely (1); unlikely (2); neutral (3); likely (4); very likely (5).

Table B.1 Variables for the beer dataset (continued).

Variable	Description
italian_food	How likely are you to eat Italian food when available? Very unlikely (1); unlikely (2); neutral (3); likely (4); very likely (5).
life_rewarding	"I feel life is very rewarding!": strongly agree (1) to strongly disagree (10).
marital_status	Single (1), in a relationship (2), cohabiting (3), married (4), divorced (5), widowed (6).
meals_dinner_friend	What would you serve a friend for dinner?
mother_education	Less than high school (1); high school degree (2); some college degree (3); college degree (4); graduate degree (5).
mother_profession	Mother's profession.
nutritional_check	Checking nutritional values frequency: Never (1); on certain products only (2); very rarely (3); on most products (4); on everything (5).
on_off_campus	Living situation: on campus (1); rent out of campus (2); live with my parents and commute (3); own my own house (4).
parents_cook	Approximately how many days a week did your parents cook? Almost everyday (1); 2–3 times a week (2); 1–2 times a week (3); on holidays only (4); never (5).
pay_meal_out	How much would you pay for a meal out? <$5.00 (1); $5.01–$10.00 (2); $10.01–$20.00 (3); $20.01–$30.00 (4); $30.01–$40.00 (5); >$40.01 (6).
persian_food	How likely are you to eat Persian food when available: very unlikely (1); unlikely (2); neutral (3); likely (4); and very likely (5).
self_perception_weight	Self-perception of weight: Slim (1); very fit (2); just right (3); slightly overweight (4); overweight (5); I don't think of myself in these terms (6).

Table B.1 Variables for the beer dataset (continued).

Variable	Description
soup	Which of the two pictures do you associate with the word soup? Veggie soup (1); creamy soup (2).
sports	Do you do any sporting activity? Yes (1); no (2); no answer (99).
thai_food	How likely are you to eat Thai food when available? Very unlikely (1); unlikely (2); neutral (3); likely (4); very likely (5).
tortilla_calories	Guess for calories in a burrito sandwich from Chipotle: 580 (1); 725 (2); 940 (3); 1165 (4).
turkey_calories	Guess for calories in Panera Bread roasted turkey and avocado BLT: 345 (1); 500 (2); 690 (3); 850 (4).
type_sports	In what type of sports are you involved?
veggies_day	How likely are you to eat veggies in a day? Very unlikely (1); unlikely (2); neutral (3); likely (4); very likely (5).
vitamins	Do you take any supplements or vitamins? Yes (1); no (2).
waffle_calories	Guess for calories in a waffle potato sandwich: 575 (1); 760 (2); 900 (3); 1315 (4).
weight	What is you weight in pounds?

Useful Mathematical Results

C.1   BRIEF OVERVIEW OF EIGENVALUES

Let $\mathbf{A}$ be an $p \times p$ matrix. Then $\lambda \in \mathbb{R}$ is an eigenvalue of $\mathbf{A}$ if there exists a non-zero vector $\mathbf{v}$ such that

$$\mathbf{A}\mathbf{v} = \lambda\mathbf{v}. \tag{C.1}$$

Now, (C.1) can also be written

$$(\mathbf{A} - \lambda\mathbf{I}_p)\mathbf{v} = 0,$$

where $\mathbf{I}_p$ is the $p \times p$ identity matrix, and so

$$|\mathbf{A} - \lambda\mathbf{I}_p| = 0.$$

Note that if $\mathbf{A}$ is a diagonal matrix with diagonal elements $a_{11}, a_{22}, \ldots, a_{pp}$, then

$$|\mathbf{A} - \lambda\mathbf{I}_p| = (a_{11} - \lambda)(a_{22} - \lambda) \times \cdots \times (a_{pp} - \lambda) = 0$$

and the eigenvalues of $\mathbf{A}$ are $a_{11}, a_{22}, \ldots, a_{pp}$.

C.2   SELECTED LINEAR ALGEBRA RESULTS

The following theorems are taken from Graybill (1983).

Theorem C.1 *Let* $\mathbf{A}$ *and* $\mathbf{B}$ *be any matrices such that* $\mathbf{AB}$ *is defined; then*

$$(\mathbf{AB})' = \mathbf{B}'\mathbf{A}'.$$

Theorem C.2 *If* $\mathbf{A}$ *is any matrix, then* $\mathbf{A}'\mathbf{A}$ *and* $\mathbf{A}\mathbf{A}'$ *are symmetric.*

Theorem C.3 *If* $\mathbf{A}$ *is a non-singular matrix, then* $\mathbf{A}'$ *and* $\mathbf{A}^{-1}$ *are non-singular and*

$$\left(\mathbf{A}'\right)^{-1} = \left(\mathbf{A}^{-1}\right)'.$$

Theorem C.4 *If each element of the* i^{th} *row of an* $n \times n$ *matrix* $\mathbf{A}$ *contains a given factor* k, *then we may write* $|\mathbf{A}| = k|\mathbf{B}|$, *where the rows of* $\mathbf{B}$ *are the same as the rows of* $\mathbf{A}$ *except that the number* k *has been factored from each element of the* i^{th} *row of* $\mathbf{A}$.

A corollary of Theorem C.4 is given below as Corollary 1, and is stated as a result by Anton and Rorres (1994).

Corollary 1 *Let* $\mathbf{A}$ *and* $\mathbf{B}$ *be* $n \times n$ *matrices such that*

$$\mathbf{A} = k\mathbf{B},$$

where k *is a scalar. Then*

$$|\mathbf{A}| = k^n|\mathbf{B}|.$$

From this corollary, it follows that, for $a \in \mathbb{R}^+$ and a $p \times p$ identity matrix $\mathbf{I}_p$,

$$\log|a^{-1}\mathbf{I}_p| = \log a^{-p} + \log|\mathbf{I}_p| = p\log a^{-1} = -p\log a.$$

The following results are taken from Chapter 4 of Lütkepohl (1996) and are also available elsewhere:

$$\mathbf{A}_{m\times n}, \mathbf{B}_{n\times m} : \ \mathrm{tr}\{\mathbf{A}\mathbf{B}\} = \mathrm{tr}\{\mathbf{B}\mathbf{A}\}.$$
$$\mathbf{A}_{m\times m}, \mathbf{B}_{m\times m} : |\ \mathbf{A}\mathbf{B}\ | = |\ \mathbf{A}\ ||\ \mathbf{B}\ |.$$
$$\mathbf{A}_{m\times m} : |\ \mathbf{A}'\ | = |\ \mathbf{A}\ |.$$
$$\mathbf{A}_{m\times m}, \text{ non-singular} : |\ \mathbf{A}^{-1}\ | = |\ \mathbf{A}\ |^{-1}.$$
$$\mathbf{A}_{m\times m} = [a_{ij}], \text{ triangular} : |\ \mathbf{A}\ | = \prod_{i=1}^{m} a_{ii}.$$

C.3 MATRIX CALCULUS RESULTS

Assume that all matrices and vectors are real, all objects that are differentiated are continuously differentiable, and all differentials are well-defined. The following results, taken from Chapter 10 of Lütkepohl (1996), are also available elsewhere:

$$\mathbf{X}_{m \times m} \text{ non-singular}: \quad \frac{\partial \log |\mathbf{X}|}{\partial \mathbf{X}} = (\mathbf{X}')^{-1}.$$

$$\mathbf{X}_{m \times n}, \mathbf{A}_{n \times m}: \quad \frac{\partial \operatorname{tr}\{\mathbf{X}\mathbf{A}\}}{\partial \mathbf{X}} = \frac{\partial \operatorname{tr}\{\mathbf{A}\mathbf{X}\}}{\partial \mathbf{X}} = \mathbf{A}'.$$

$$\mathbf{X}_{m \times n}, \mathbf{A}_{m \times n}: \quad \frac{\partial \operatorname{tr}\{\mathbf{X}'\mathbf{A}\}}{\partial \mathbf{X}} = \frac{\partial \operatorname{tr}\{\mathbf{A}\mathbf{X}'\}}{\partial \mathbf{X}} = \mathbf{A}.$$

$$\mathbf{X}_{m \times n}, \mathbf{A}_{p \times m}, \mathbf{B}_{n \times p}: \quad \frac{\partial \operatorname{tr}\{\mathbf{A}\mathbf{X}\mathbf{B}\}}{\partial \mathbf{X}} = \mathbf{A}'\mathbf{B}'.$$

$$\mathbf{X}_{m \times n}, \mathbf{A}_{n \times n} \text{ symmetric}: \quad \frac{\partial \operatorname{tr}\{\mathbf{X}\mathbf{A}\mathbf{X}'\}}{\partial \mathbf{X}} = 2\mathbf{X}\mathbf{A}.$$

$$\mathbf{X}_{m \times n}, \mathbf{A}_{n \times m}, \mathbf{B}_{n \times m}: \quad \frac{\partial \operatorname{tr}\{\mathbf{X}\mathbf{A}\mathbf{X}\mathbf{B}\}}{\partial \mathbf{X}} = \mathbf{B}'\mathbf{X}'\mathbf{A}' + \mathbf{A}'\mathbf{X}'\mathbf{B}'.$$

$$\mathbf{X}_{m \times n}, \mathbf{A}_{n \times n}, \mathbf{B}_{m \times m}: \quad \frac{\partial \operatorname{tr}\{\mathbf{X}\mathbf{A}\mathbf{X}'\mathbf{B}\}}{\partial \mathbf{X}} = \mathbf{B}'\mathbf{X}\mathbf{A}' + \mathbf{B}\mathbf{X}\mathbf{A}.$$

$$\mathbf{X}_{m \times n}, \mathbf{A}_{p \times m}, \mathbf{B}_{m \times p}: \quad \frac{\partial \operatorname{tr}\{\mathbf{A}\mathbf{X}\mathbf{X}'\mathbf{B}\}}{\partial \mathbf{X}} = (\mathbf{B}\mathbf{A} + \mathbf{A}'\mathbf{B}')\mathbf{X}.$$

$$\mathbf{X}_{n \times n} \text{ non-singular}: \quad \frac{\partial |\mathbf{X}^{-1}|}{\partial \mathbf{X}} = -|\mathbf{X}|^{-1}(\mathbf{X}')^{-1}.$$

A result related to the latter is also useful:

$$\mathbf{X}_{n \times n} \text{ non-singular}: \quad \frac{\partial |\mathbf{X}|}{\partial \mathbf{X}^{-1}} = -|\mathbf{X}|\mathbf{X}'.$$

Performance Tips

The material in this appendix is based on the Julia performance tips detailed in the Julia manual[1] and the material in Goldberg (1991).

D.1 FLOATING POINT NUMBERS

D.1.1 Do Not Test for Equality

When comparing floating point numbers, check that the absolute value of their difference is less than some tolerance. Floating point operations often involve a tiny loss of precision, making two numbers unequal at the bit level when they are actually equal enough for practical purposes.

```
a = 3.1415926
tol = 1e-5

## Bad
if a == pi
  println("a is equal to $pi")
end

## Good
if abs(a - pi) < tol
  println("a is equal to $pi")
end

## Better
if isapprox(a, pi, rtol = tol)
  println("a is equal to $pi")
end
```

---

[1]docs.julialang.org/en/stable/manual/performance-tips/

D.1.2   Use Logarithms for Division

When doing division with large numbers, it can be advantageous to replace them with their logarithms. The following code block shows how two large numbers, generated with the `gamma()` function, can be divided using their logarithms.

```
using SpecialFunctions

## Two very large numbers
/(gamma(210), gamma(190))
#NaN

exp(-(lgamma(210), lgamma(190)))
#9.89070132023e45
```

D.1.3   Subtracting Two Nearly Equal Numbers

If two numbers agree to b bits, b bits of precision can be lost if they are subtracted. If the two numbers have identical machine representations, their difference could be zero, despite this being mathematically false. A simple way to illustrate this is taking the limit of a function at a particular point. The following code block shows what happens when we take the limit of $\exp(1)$. As $h \to 0$, the limit reaches the true value at $h = 10^{-5}$ and remains there until $h = 10^{-11}$. At $h = 10^{-16}$, the two numbers in the numerator of the limit expression are equal to machine precision and the result is zero. When this occurs, it is often beneficial to reformulate the problem so that it does not require a subtraction.

```
## limit of exp(1)
lim_exp1(h) = /( -(exp(1. + h), exp(1.)), h )

## h from 10e-4 to 10e-16
for i in 4:16
  le1 = lim_exp1(10.0^(-i))
  println("$i : ", le1)
  println("derivative == exp(1): $(isapprox( exp(1), le1, rtol = tol ))")
end

# 4 : 2.718417747082924
# derivative == exp(1): false
# 5 : 2.7182954199567173
# derivative == exp(1): true
# ..
# 10 : 2.7182833761685288
# derivative == exp(1): true
# 11 : 2.7183144624132183
# derivative == exp(1): false
# ..
# 16 : 0.0
# derivative == exp(1): false
```

D.2 JULIA PERFORMANCE

D.2.1 General Tips

The following three tips will almost without exception improve your Julia code:

1. Do not use global variables.

2. Write functions.

3. Profile your code.

The first tip is important because the Julia compiler has a hard time optimizing global variables because their type can change at any point in the program. If they must be used, they should be defined as constants using the `const` keyword. Whenever possible, constants should have a type associated with them.

The second tip is there because of the way the compiler works; specifically, code inside functions is typically much faster. Functions have well-defined arguments and outputs which help the compiler make optimizations. Additionally, functions are helpful for the maintenance and testing of your software, both very important aspects of software development.

The third tip is crucial as it can help data scientists discover problems and improve the performance of code. The Julia package ecosystem has a few options for doing profiling, such as the profile module `ProfileView.jl` and `BenchmarkTools.jl`. We will briefly touch on the `time` macro, which is used in the following code block (Appendix D.2.2), where `@time` is added before a Julia expression and returns the time the expression took to execute, the number of allocations it made, and the total number of bytes the allocations used. Large memory allocation is often an indication of a problem in code. Memory allocation problems are often related to type instability or not using mutable data structures properly. Note that the first time `@time` is run, it is being compiled and so this timing should be ignored. Initially, we run `@time` two to three times before we start to take note of the information it provides.

D.2.2 Array Processing

In Julia, similar to R, arrays are stored in column major order. When processing two-dimensional arrays, it is much faster to process them by first iterating over the columns and then the rows.

The following code block compares two functions that total the entries in a 10000×3000 array of random numbers. The function that uses an outer loop for the columns is roughly 70% faster.

```
using Random
Random.seed!(63)

## Processing arrays
A1 = rand(Float64, (10000,3000))

function outer_row(A)
  tot = 0
  m,n = size(A)
  for i = 1:m, j = 1:n
    tot += A[i, j]
  end
  return tot
end

function outer_col(A)
  tot = 0
  m,n = size(A)
  for j = 1:n, i = 1:m
    tot += A[i, j]
  end
  return tot
end

@time outer_row(A1)
# 0.725792 seconds (5 allocations: 176 bytes)

@time outer_col(A1)
# 0.207118 seconds (5 allocations: 176 bytes)
```

When writing functions that return an array, it is often advantageous to pre-allocate the memory for the return value. This will cut down on performance bottlenecks associated with memory allocation and garbage collection. Pre-allocation also has the advantage of providing some type control on the object being returned. The following code block details how one might take the mean of each column in an array. This is just for illustration and ignores the fact that the mean() function can complete the task by specifying the correct dimension in its arguments.

The first function colm() is very inefficient. The returned array is defined at the beginning as the res object. Because no type is specified, Julia assumes it can take entries of any type, making for poor optimizations. At each value of j, colm() makes a copy of the array slice, calculates its mean and concatenates it to the result array res. This results in a function call that does approximately 91,000 allocations and uses 267 MB of memory.

The second function `colm2()` allocates a 3000-element array of type `Float64` to store the column means, as we know the number of columns in the input. Inside the loop, it takes advantage of array views, which returns a `view` into the array at the specified indices without making a copy of the slice. These modifications result in an 80% speed improvement and 97% fewer memory allocations.

```
using Random
Random.seed!(65)

## Processing arrays
A1 = rand(Float64, (10000,3000))

## Inefficient
function colm(A)
  res = Vector()
  m,n = size(A)
  for j = 1:n
    res = vcat(res, mean(A[:, j]))
  end
  return res
end

@time colm(A1) # Any[3000]
#  0.561565 seconds (90.89 k allocations: 266.552 MiB, 21.93% gc time)

## efficient
function colm2(A)
  m,n = size(A)
  res = Vector{Float64}(undef, n)
  for j = 1:n
    res[j] = mean(view(A, :, j))
  end
  return res
end

@time colm2(A1) # Float64[3000]
# 0.118919 seconds (3.01 k allocations: 164.297 KiB
```

D.2.3 Separate Core Computations

When writing Julia functions, it is recommended that complex and/or repeated computations be implemented in separate functions. This can help the compiler optimize the code and aids with debugging efforts. Another benefit is the increased opportunity for code re-use.

The concept is illustrated in the next code block. The first function `matvec()` does matrix-vector multiplication by expressing the result as a linear combination of the matrix's columns, where the coefficients are the vector's entries. The function does this in a

for loop and returns the result as a dictionary. The function has
two components, the first dealing with the dictionary initialization
and population and the second concerning the multiplication. If we
separate these two components into separate functions, mvmul()
and matvec2(), we gain in performance with a 24% speed up, and
also have more readable and maintainable code.

```
## sample matrix and vector
M1 = [1 2 3; 2 4 6 ] #2x3
V1 = [1,2,2] # 3x1

## column representation of the multiplication
function matvec(M, v)
    d1 = Dict{String, Vector{Real}}()
    m,n = size(M)
    res2 = zeros(m)
    for i = 1:n
        res2 = +(res2, *(view(M, :, i), v[i]))
    end
    return push!(d1, "M1xV1"=>res2)
end

@time matvec(M1, V1)
# 0.000021 seconds (21 allocations: 1.672 KiB)

## separate function to do the computation
## use dispatch so it accepts Int and Float arguments
function mvmul(M::Matrix{T}, v::Vector{T}) where {T <: Number}
    m,n = size(M)
    res2 = zeros(m)
    for i = 1:n
        res2 = +(res2, *(view(M, :, i), v[i]))
    end
    return res2
end

## calls the computation function
function matvec2(M, v)
    d1 = Dict{String, Vector{Real}}()
    v1 = mvmul(M, v)
    return push!(d1, "M1xV1"=>v1)
end

@time matvec2(M1, V1)
# 0.000016 seconds (21 allocations: 1.672 KiB)
```

Linear Algebra Functions

This appendix will help readers translate the linear algebra they know into Julia code. This material is based on the official documentation for the LinearAlgebra[1] package and Eldén (2007).

E.1 VECTOR OPERATIONS

Some common vector operations in Julia are given in Table E.1.

Note that the Julia operations given in Table E.1 should not be considered an exclusive list. For example, both dot(x,y) and x·y are quoted for the inner product but ·(x,y) also works. The following code block illustrates some of the operations in Table E.1.

```
x = [1, 2]
y = [3, 4]
c = 8

c*x
# 2-element Array{Int64,1}:
#  8
#  16

dot(x,y)
# 11

x + c*ones(2)
# 2-element Array{Float64,1}:
#   9.0
#  10.0
```

---

[1]docs.julialang.org/en/v1/stdlib/LinearAlgebra/index.html

```
normalize(x)
# 2-element Array{Float64,1}:
#  0.4472135954999579
#  0.8944271909999159
```

Table E.1 Common vector operations in Julia, using LinearAlgebra, for n-dimensional vectors **x** and **y**.

Operation	Notation	Julia
Addition	$\mathbf{x} + \mathbf{y}$	x+y or +(x, y)
Subtraction	$\mathbf{x} - \mathbf{y}$	x-y or -(x, y)
Scalar-vector addition	$\mathbf{x} + c\mathbf{1}$	x+c*ones(n)
Scalar-vector multiplication	$c\mathbf{x}$	c*x or *(c, x)
Transpose	$\mathbf{x}'$ or $\mathbf{x}^\top$	x' or transpose(x)
Inner product	$\mathbf{x}'\mathbf{y}$	dot(x,y) or x·y
Cross product	$\mathbf{x} \times \mathbf{y}$	cross(x,y) or x×y
Norm	$\| \mathbf{x} \|$	norm(x)
Distance	$\| \mathbf{x} - \mathbf{y} \|$	norm(x-y)
Normalize	$\mathbf{x}/\| \mathbf{x} \|$	normalize(x)
Sum	$\sum_{i=1}^{n} x_i$	sum(x)
Mean	$(1/n) \sum_{i=1}^{n} x_i$	mean(x)

E.2 MATRIX OPERATIONS

Some common matrix operations in Julia are given in Table E.2; again, this is not intended to be an exhaustive list.

The following code block illustrates some of the operations in Table E.2.

```
X = [5 1; 2 4]
Y = [2 0; 1 5]
y = [3, 4]

X*y
# 2-element Array{Int64,1}:
#  19
#  22

*(Y,X)
# 2x2 Array{Int64,2}:
#  10   2
#  15  21

## Row sums
```

```
sum(X,dims=1)
# 1x2 Array{Int64,2}:
#  7  5
```

Table E.2 Common matrix operations in Julia, where all operations assume compatible matrix and vector dimensions.

Operation	Notation	Julia
Addition	$\mathbf{X} + \mathbf{Y}$	X+Y or +(X,Y)
Subtraction	$\mathbf{X} - \mathbf{Y}$	X-Y or -(X,Y)
Transpose	$\mathbf{X}'$ or $\mathbf{X}^\top$	X' or transpose(X)
Inverse	$\mathbf{X}^{-1}$	inv(X)
Moore-Penrose pseudo-inv.	$\mathbf{X}^\dagger$	pinv(X)
Scalar-matrix multiplication	$c\mathbf{X}, c \in \mathbb{R}$	*(c,X) or c*X
Vector-matrix multiplication	$\mathbf{X}\mathbf{y}$	*(X,y) or X*y
Matrix-matrix multiplication	$\mathbf{X}\mathbf{Y}$	*(X,Y) or X*Y
Matrix raised to a power p	$\mathbf{X}^p$	X^p
Determinant	$\mid \mathbf{X} \mid$	det(X)
Log-determinant	$\log(\mid\mathbf{X}\mid)$	logdet(X)
Log absolute value det.	$\log(\mid\mid\mathbf{X}\mid\mid)$	logabsdet(X)
Column sum	$\sum_{m \in M} x_{m,n}$	sum(X, dims=1)
Row sum	$\sum_{n \in N} x_{m,n}$	sum(X, dims=2)

E.3 MATRIX DECOMPOSITIONS

Some common matrix decompositions in Julia are given in Table E.3.

The following code block illustrates how to use the decompositions in Table E.3.

```
using LinearAlgebra

X = [5 1 3; 0 8 2; 3 1 6]
Y = [8 0 1; 0 3 2; 1 2 5] # symmetric positive-definite

## Eigenvalue decomposition
evX=eigen(X)

# eigenvectors
evX.vectors
# 3x3 Array{Float64,2}:
#  0.7024    -0.384529   -0.497977
#  0.242773   0.7843     -0.652046
# -0.6691    -0.486837   -0.571712
```

```
# eigenvalues
evX.values
# 3-element Array{Float64,1}:
#   2.487860053322798
#   6.758543755579095
#   9.753596191098106

## Singular value decomposition
svdX=svd(X)

# matrix U
svdX.U
# 3x3 Array{Float64,2}:
#  -0.476574   0.457265  -0.750857
#  -0.655574  -0.753904  -0.0430237
#  -0.585747   0.471739   0.659062

# matrix V
svdX.V
# 3x3 Adjoint{Float64,Array{Float64,2}}:
#  -0.422445   0.539959  -0.728
#  -0.643539  -0.744284  -0.178604
#  -0.638278   0.393046   0.661904

# matrix Sigma
diagm(0 => svdX.S)
# 3x3 Array{Float64,2}:
#  9.80036  0.0      0.0
#  0.0      6.85522  0.0
#  0.0      0.0      2.44107

## Cholesky decomposition
cholY=cholesky(Y)

# matrix L
cholY.L
# 3x3 LowerTriangular{Float64,Array{Float64,2}}:
#  2.82843    .        .
#  0.0       1.73205   .
#  0.353553  1.1547   1.88193

# note that cholY.U gives transpose(cholY.L)

## QR decomposition
qrX = qr(X)

# matrix Q
qrX.Q
# 3x3 LinearAlgebra.QRCompactWYQ{Float64,Array{Float64,2}}:
#  -0.857493   0.0220386  -0.514024
#   0.0       -0.999082   -0.0428353
#  -0.514496  -0.036731    0.856706

#matrix R
qrX.R
# 3x3 Array{Float64,2}:
#  -5.83095  -1.37199  -5.65945
#   0.0      -8.00735  -2.15243
#   0.0       0.0       3.51249
```

Table E.3 Common matrix decompositions in Julia, where Notation is intended to help in understanding the subsequent code block.

Decomposition	Notation	Julia
Eigenvalue	$\mathbf{X}_{m \times m} = \mathbf{PDP}^{-1}$	eigen(X)
Singular value	$\mathbf{X}_{m \times n} = \mathbf{U}_{m \times m} \mathbf{\Sigma}_{m \times n} \mathbf{V}'_{n \times n},$ $m \geq n$	svd(X)
Cholesky	$\mathbf{X}_{m \times m} = \mathbf{LL}'$, $\mathbf{X}$ positive definite	cholesky(X)
QR	$\mathbf{X}_{m \times n} = \mathbf{Q}_{m \times m} \mathbf{R}_{m \times n}, m \geq n$	qr(X)

References

Aitken, A. C. (1926). A series formula for the roots of algebraic and transcendental equations. *Proceedings of the Royal Society of Edinburgh 45*, 14–22.

Anderson, E. (1935). The irises of the Gaspé Peninsula. *Bulletin of the American Iris Society 59*, 2–5.

Andrews, J. L. and P. D. McNicholas (2013). *vscc: Variable Selection for Clustering and Classification*. R package version 0.2.

Andrews, J. L. and P. D. McNicholas (2014). Variable selection for clustering and classification. *Journal of Classification 31*(2), 136–153.

Anton, H. and C. Rorres (1994). *Elementary Linear Algebra* (7th ed.). New York: John Wiley & Sons.

Bezansony, J., A. Edelmanz, S. Karpinskix, and V. B. Shahy (2017). Julia: A fresh approach to numerical computing. *SIAM Review 59*(1), 65–98.

Bishop, C. M. (2006). *Pattern Recognition and Machine Learning*. New York: Springer.

Böhning, D., E. Dietz, R. Schaub, P. Schlattmann, and B. Lindsay (1994). The distribution of the likelihood ratio for mixtures of densities from the one-parameter exponential family. *Annals of the Institute of Statistical Mathematics 46*, 373–388.

Box, D. and A. Hejlsberg (2007). LINQ: .NET language-integrated query. msdn.microsoft.com/en-us/library/bb308959.aspx.

Breiman, L. (1996). Bagging predictors. *Machine Learning 24*(2), 123–140.

Breiman, L. (2001a). Random forests. *Machine Learning 45*(1), 5–32.

Breiman, L. (2001b). Statistical modeling: The two cultures. *Statistical Science 16*(3), 199–231.

Breiman, L., J. H. Friedman, R. A. Olshen, and C. J. Stone (1984). *Classification and Regression Trees.* Boca Raton: Chapman & Hall/CRC Press.

Brier, G. W. (1950). Verification of forecasts expressed in terms of probability. *Monthly Weather Review 78*(1), 1–3.

Briggs, D. E., C. A. Boulton, P. A. Brookes, and R. Stevens (2004). *Brewing: Science and Practice.* Boca Raton: CRC Press.

Browne, R. P. and P. D. McNicholas (2014). *mixture: Mixture Models for Clustering and Classification.* R package version 1.1.

Carr, D. B., R. J. Littlefield, W. Nicholson, and J. Littlefield (1987). Scatterplot matrix techniques for large n. *Journal of the American Statistical Association 82*(398), 424–436.

Chen, T. and C. Guestrin (2016). XGBoost: a scalable tree boosting system. In *Proceedings of the 22nd ACM SIGKDD international conference on knowledge discovery and data mining,* San Francisco, pp. 785–794. ACM.

Cleveland, W. S. (2001). Data science: An action plan for expanding the technical areas of the field of statistics. *International Statistical Review/Revue Internationale De Statistique 69*(1), 21–26.

Davenport, T. H. and D. J. Patil (2012). Data scientist: The sexiest job of the 21st century. *Harvard Business Review.* Sourced from hbr.org/2012/10/data-scientist-the-sexiest-job-of-the-21st-century

Davison, A. C. and D. V. Hinkley (1997). *Bootstrap Methods and their Application.* New York: Cambridge University Press.

Dempster, A. P., N. M. Laird, and D. B. Rubin (1977). Maximum likelihood from incomplete data via the EM algorithm. *Journal of the Royal Statistical Society: Series B 39*(1), 1–38.

Efron, B. (1979). Bootstrap methods: Another look at the jack-knife. *The Annals of Statistics* 7(1), 1–26.

Efron, B. (2002). *The bootstrap in modern statistics*, pp. 326–332. Statistics in the 21st Century. Boca Raton: Chapman & Hall/CRC Press.

Efron, B. and T. Hastie (2016). *Computer Age Statistical Inference*. Cambridge: Cambridge University Press.

Efron, B. and R. J. Tibshirani (1993). *An Introduction to the Bootstrap*. Boca Raton: Chapman & Hall/CRC Press.

Eldén, L. (2007). *Matrix Methods in Data Mining and Pattern Recognition*. Philadelphia: SIAM.

Everitt, B. S., S. Landau, M. Leese, and D. Stahl (2011). *Cluster Analysis* (5th ed.). Chichester: John Wiley & Sons.

Fernández-Delgado, M., E. Cernadas, S. Barro, and D. Amorim (2014). Do we need hundreds of classifiers to solve real world classification problems? *Journal of Machine Learning Research* 15(1), 3133–3181.

Friedl, J. E. F. (2006). *Mastering Regular Expressions: Understand Your Data and Be More Productive* (3rd ed.). Sebastopol, California: O'Reilly Media, Inc.

Friedman, J. H. (2001). Greedy function approximation: A gradient boosting machine. *The Annals of Statistics* 29(5), 1189–1232.

Fujikoshi, Y., V. V. Ulyanov, and R. Shimizu (2010). *Multivariate Statistics: High-Dimensional and Large-Sample Approximations*. Hoboken: John Wiley & Sons Inc.

Gelman, A., C. Pasarica, and R. Dodhia (2002). Let's practice what we preach: turning tables into graphs. *The American Statistician* 56(2), 121–130.

Geurts, P., D. Ernst, and L. Wehenkel (2006). Extremely randomized trees. *Machine Learning* 63(1), 3–42.

Ghahramani, Z. and G. E. Hinton (1997). The EM algorithm for factor analyzers. Technical Report CRG-TR-96-1, University of Toronto, Canada.

Gneiting, T. and A. E. Raftery (2007). Strictly proper scoring rules, prediction, and estimation. *Journal of the American Statistical Association 102*(477), 359–378.

Goldberg, D. (1991). What every computer scientist should know about floating-point arithmetic. *ACM Computing Surveys 23*(1), 5–48.

Graybill, F. A. (1983). *Matrices with Applications in Statistics* (2nd ed.). Belmont, California: Wadsworth.

Hardin, J., R. Hoerl, N. J. Horton, D. Nolan, B. Baumer, O. Hall-Holt, P. Murrell, R. Peng, P. Roback, D. T. Lang, and M. D. Ward (2015). Data science in statistics curricula: Preparing students to "think with data". *The American Statistician 69*(4), 343–353.

Hastie, T., R. Tibshirani, and J. Friedman (2009). *The Elements of Statistical Learning* (2nd ed.). New York: Springer.

Hayashi, C. (1998). What is data science? Fundamental concepts and a heuristic example. In C. Hayashi, K. Yajima, H. H. Bock, N. Ohsumi, Y. Tanaka, and Y. Baba (Eds.), *Data Science, Classification, and Related Methods. Studies in Classification, Data Analysis, and Knowledge Organization.* Tokyo: Springer.

Higham, N. J. (2002). *Accuracy and Stability of Numerical Algorithms* (2nd ed.). Philadelphia: SIAM.

Hintze, J. L. and R. D. Nelson (1998). Violin plots: A box plot-density trace synergism. *The American Statistician 52*(2), 181–184.

Hunter, D. R. and K. Lange (2000). Rejoinder to discussion of "Optimization transfer using surrogate objective functions". *Journal of Computational and Graphical Statistics 9*, 52–59.

Hunter, D. R. and K. Lange (2004). A tutorial on MM algorithms. *The American Statistician 58*(1), 30–37.

Kuhn, M. (2017). *caret: Classification and Regression Training.* R package version 6.0-78.

Kuhn, M. and K. Johnson (2013). *Applied Predictive Modeling.* New York: Springer-Verlag.

Lawley, D. N. and A. E. Maxwell (1962). Factor analysis as a statistical method. *Journal of the Royal Statistical Society: Series D 12*(3), 209–229.

Lindsay, B. G. (1995). Mixture models: Theory, geometry and applications. In *NSF-CBMS Regional Conference Series in Probability and Statistics*, Volume 5. California: Institute of Mathematical Statistics: Hayward.

Lopes, H. F. and M. West (2004). Bayesian model assessment in factor analysis. *Statistica Sinica 14*, 41–67.

Lütkepohl, H. (1996). *Handbook of Matrices*. Chicester: John Wiley & Sons.

McCullagh, P. and J. A. Nelder (1989). *Generalized Linear Models*. Boca Raton: Chapman & Hall/CRC Press.

McLachlan, G. J. and T. Krishnan (2008). *The EM Algorithm and Extensions* (2nd ed.). New York: Wiley.

McLachlan, G. J. and D. Peel (2000). Mixtures of factor analyzers. In *Proceedings of the Seventh International Conference on Machine Learning*, pp. 599–606. San Francisco: Morgan Kaufmann.

McNicholas, P. D. (2010). Model-based classification using latent Gaussian mixture models. *Journal of Statistical Planning and Inference 140*(5), 1175–1181.

McNicholas, P. D. (2016a). *Mixture Model-Based Classification*. Boca Raton: Chapman & Hall/CRC Press.

McNicholas, P. D. (2016b). Model-based clustering. *Journal of Classification 33*(3), 331–373.

McNicholas, P. D., A. ElSherbiny, A. F. McDaid, and T. B. Murphy (2018). *pgmm: Parsimonious Gaussian Mixture Models*. R package version 1.2.2.

McNicholas, P. D. and T. B. Murphy (2008). Parsimonious Gaussian mixture models. *Statistics and Computing 18*(3), 285–296.

McNicholas, P. D. and T. B. Murphy (2010). Model-based clustering of microarray expression data via latent Gaussian mixture models. *Bioinformatics 26*(21), 2705–2712.

McNicholas, P. D., T. B. Murphy, A. F. McDaid, and D. Frost (2010). Serial and parallel implementations of model-based clustering via parsimonious Gaussian mixture models. *Computational Statistics and Data Analysis 54*(3), 711–723.

Meng, X.-L. and D. B. Rubin (1993). Maximum likelihood estimation via the ECM algorithm: a general framework. *Biometrika 80*, 267–278.

Meng, X.-L. and D. van Dyk (1997). The EM algorithm — an old folk song sung to a fast new tune (with discussion). *Journal of the Royal Statistical Society: Series B 59*(3), 511–567.

Oliver, G. and T. Colicchio (2011). *The Oxford Companion to Beer.* Oxford University Press.

Press, G. (2013). A very short history of data science. Sourced from `www.forbes.com/sites/gilpress/2013/05/28/a-very-short-history-of-data-science/`.

Puts, M., P. Daas, and T. de Waal (2015). Finding errors in big data. *Significance 12*(3), 26–29.

R Core Team (2018). *R: A Language and Environment for Statistical Computing.* Vienna, Austria: R Foundation for Statistical Computing.

Ridgeway, G. (2017). *gbm: Generalized Boosted Regression Models.* With contributions from others. R package version 2.1.3.

Ruppert, D., M. P. Wand, and R. J. Carroll (2003). *Semiparametric Regression.* Cambridge Series in Statistical and Probabilistic Mathematics. Cambridge: Cambridge University Press.

Schutt, R. (2013). *Doing Data Science.* Sebastopol, California: O'Reilly Media, Inc.

Schwarz, G. (1978). Estimating the dimension of a model. *The Annals of Statistics 6*(2), 461–464.

Spearman, C. (1904). The proof and measurement of association between two things. *American Journal of Psychology 15*, 72–101.

Spearman, C. (1927). *The Abilities of Man: Their Nature and Measurement.* London: MacMillan and Co., Limited.

Streuli, H. (1973). Der heutige stand der kaffeechemie. In *Association Scientifique International du Cafe, 6th International Colloquium on Coffee Chemistry*, Bogatá, Colombia, pp. 61–72.

Tipping, M. E. and C. M. Bishop (1999a). Mixtures of probabilistic principal component analysers. *Neural Computation 11*(2), 443–482.

Tipping, M. E. and C. M. Bishop (1999b). Probabilistic principal component analysis. *Journal of the Royal Statistical Society. Series B 61*, 611–622.

Tukey, J. W. (1962). The future of data analysis. *The Annals of Mathematical Statistics 33*(1), 1–67.

Tukey, J. W. (1977). *Exploratory Data Analysis*. Reading, Massachusetts: Addison-Wesley.

van Rossum, G. (1995). Python reference manual. Centrum voor Wiskunde en Informatica (CWI) Report CS-R9525. CWI: Amsterdam, The Netherlands.

Venables, W. N. and B. D. Ripley (2002). *Modern Applied Statistics with S* (4th ed.). New York: Springer.

White, T. (2015). *Hadoop: The Definitive Guide* (4th ed.). Sebastopol, California: O'Reilly Media, Inc.

Wickham, H. (2009). *ggplot2: Elegant Graphics for Data Analysis*. New York: Springer.

Wickham, H. (2011). The split-apply-combine strategy for data analysis. *Journal of Statistical Software 40*(1), 1–29.

Wickham, H. (2016). *plyr: Tools for Splitting, Applying and Combining Data*. R package version 1.8.4.

Wickham, H., R. Francois, L. Henry, and K. Müller (2017). *dplyr: A Grammar of Data Manipulation*. R package version 0.7.4.

Wilkinson, L. (2005). *The Grammar of Graphics* (2nd ed.). New York: Springer-Verlag.

Woodbury, M. A. (1950). *Inverting modified matrices*. Statistical Research Group, Memorandum Report 42. Princeton, New Jersey: Princeton University.

Wright, M. N. and A. Ziegler (2017). ranger: A fast implementation of random forests for high dimensional data in C++ and R. *Journal of Statistical Software 77*(1), 1–17.

Zuras, D., M. Cowlishaw, A. Aiken, M. Applegate, D. Bailey, S. Bass, D. Bhandarkar, M. Bhat, D. Bindel, S. Boldo, et al. (2008). IEEE standard for floating-point arithmetic. *IEEE Std 754-2008*, 1–70.

Index

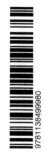